Acknowledgements

Even in writing a modest book such as this the author owes a debt for the ideas, facts, comments and patience provided by many people. I would therefore like to thank my own teachers for awakening my interest in physical geography (especially David Ingle Smith) and my colleagues for encouraging it (especially Vincent Tidswell). My students and pupils over the years have also provided an invaluable source of comment, criticism and inspiration. Of crucial importance, however, has been the support of my wife Alison who has endured the eighteen months of writing and production with endless patience. In connection with the production of the book I owe a special debt to Chris Kington for his early editorial work, Vincent Driver for the diagrams, and Denys Brunsden and Peta Hambling for help in locating photographs. For the remaining errors and omissions I alone remain responsible.

Keith Hilton
Kew, 1979

Sources of illustrations

Although all diagrams and maps have been drawn especially for this book the author would like to acknowledge the following sources for the data and/or ideas:

Andrews J T, *IBG Spec. Pub. 2 (1970)*, **6.26**; Barr W, *Arctic (24, 4/1971)*, **6.25**; Barry R G & Chorley R, *Atmosphere Weather & Climate*, Methuen, **2.3, 2.13, 2.18, 2.24, 2.30**; Bloom A L, *Geomorphology*, Prentice Hall, **1.5**; Brunsden D, *IBG Spec. Pub. 7 (1974)*, **3.18, 3.20**; Brunsden D & Doornkamp J C, *The Unquiet Landscape*, David & Charles, **1.4, 3.30, 4.27**; Brunsden D & Kesel R H, *J. Geol. (81/1973)*, **3.37**; Carlston C W, *USGS Prof. Paper 422c (1963)*, **4.44**; Carson M & Petley D, *IBG Trans. (49/1970)*, **3.33**; Chandler T, *The Climate of London*, Hutchinson, **7.8**; Chandler T J & Musk L F, *Geog. Mag. (11/1976)*, **2.26**; Chorley R J, *Water, Earth & Man*, Methuen, **2.11, 2.12, 2.14, 4.4**; Chorley R J & Kennedy B, *Physical Geography*, Prentice Hall, **3.9**; Chorley R J & Morgan M A, *Geol. Soc. Am. Bull. (1962)*, **4.41**; Chow Ven Te *Handbook of Applied Hydrology*, McGraw Hill, **4.78, 4.79**; Courtney F & Trudgill S, *The Soil*, Arnold, **5.27**; Elliot P, *Weather (19/1964)*, **2.19A**; Fairbridge R, *The Encyclopedia of Geomorphology*, Reinhold, **3.24, 4.23, 4.25, 4.29, 6.2**; Flint R F, *Glacial & Quaternary Geology*, Wiley, **6.27, 6.42**; Flohn H, *Climate & Weather*, Weidenfeld & Nicolson, **2.8, 2.21**; Gass I, Smith A & Wilson R, *Understanding the Earth*, Open Univ, **1.6, 1.7, 1.8, 1.9**; Geiger R, *Climate near the Ground*, Harvard, **2.9**; Ghersmehl P J, *AAAG (66, 2/1976)*, **5.13, 5.14, 5.19, 5.23**; Geology Survey of Canada, **6.18, 6.19, 6.26**; Gregory K J & Walling D E, *Drainage Basin Form & Process*, Arnold, **4.8, 4.10, 4.19, 4.21, 4.29, 4.30, 4.33, 4.45, 4.77**; Hanwell J G & Newson M D, *Wessex Cave Club Occ. Paper 2 (1970)*, **4.49**; *HMSO Surface Water Year Book 1965–6*, **4.15, 4.61**; Howe G, Slaymaker H O & Harding D, *IBG Trans. (41/1967)*, **4.44**; Ives J D & Barry R G, *Arctic & Alpine Environments*, Methuen, **6.4, 6.36**; Jones D, *Geog. Mag. (12/1974)*, **1.13**; Leopold L, Wolman M G & Miller J P, *Fluvial Processes in Geomorphology*, Freeman, **4.29, 4.35, 4.46, 4.48, 4.57**; Loken O & Hodgson D, *Can. J. Earth Sci. (8/1971)*, **6.29**; Mason B J, *QJR Met. Soc. (96/1970)*, **2.37**; Meier M F, *USGS Prof. Paper 351 (1960)*, **6.5**; Murray R, *Met. Mag. (106/1977)*, **2.29**; *National Geographic Society Magazine 01/1973*: **1.12, 1.15, 1.16, 1.19, 11/1976: 6.1, 6.2, 7.1, 7.3**; Newson M D, *IBG Trans. (54/1971)*, **3.14**; Rahn P H, *J. Geol. (19/1971)*, **3.1**; Rapp A, *Geog. Annaler (42/1961)*, **3.29**; Ratcliffe R, *Met. Mag. (106/1977)*, **2.29**; Richards J H, *The Atlas of Saskatchewan*, Univ. Saskatchewan, **4.68, 4.69, 4.71, 4.72, 4.73, 5.18**; Rodda J C, *IBG Trans. (49/1970)*, **4.9, 4.75**; Ruhe R V, *Am. J. Sci. (250/1952)*, **4.47**; Scovill J L, *Atlas of Landforms*, Wiley, **3.35, 6.21, 6.27**; Sellers W D, *Physical Climatology*, Chicago, **2.2, 2.6, 2.27, 5.9**; Simpson J E (et al.), *QJR Met. Soc. (103/1977)*, **2.19**; Sissons J B, *IBG Trans. (2/1977)*, **6.10**; Sissons J B, *Scotland*, Methuen, **6.31**; Soil Survey of Great Britain, **5.24**; Smith D I, *Limestones and Caves of the Mendip Hills*, David & Charles, **3.11, 3.12, 4.12**; Smith D I & Mead D, *Proc. Univ. Bristol Spel. Soc. (1962)*, **3.13**; Strahler A N, *Planet Earth: Its Geological Systems through Geologic Time*, Harper & Row, **2.7, 2.16, 2.33, 3.2**; Statham I, *IBG Trans. (59/1973)*, **3.8**; Thornes J B & Brunsden D, *Geomorphology & Time*, Methuen, **3.19**; Weyman D, *Runoff Processes & Streamflow Modelling*, OUP, **4.13**; Wye River Authority (WNWDA) **4.61, 4.62, 4.63, 4.64, 4.65, 4.66**; Yudkovitch N, *Albertan Geographer (3/1967)*, **7.5, 7.6**.

Photographs were supplied by the following:

Aerofilms Ltd **4.53, 5.10, 6.11**; Prof W Barr **4.70, 6.24**; Camera Press **5.5**; Canadian Government DREE **4.74**, NFB **7.2**, RCAF **5.17, 6.20**; Fairey Surveys Ltd **3.17, 4.32**; reproduced by permission of the Director, Institute of Geological Sciences; crown copyright reserved; **3.4, 3.6, 3.15, 3.34, 6.9**; Geoslides **5.8**; Prof J D Ives **6.7**; F W Lane **1.10**; NASA **2.32, 4.1, 4.2, 6.35** and **cover**; Popperfoto **4.22**; Dr J S Shelton **3.23, 4.54, 4.55, 4.59**; US Army Corps of Engineers **4.26**; US Forest Service **3.22**.

Contents

Introduction: the Scope and Purpose of the Book

The processes and patterns which form the theme of this book are not merely those concerned with the earth's surface. They also include the *processes of enquiry and explanation*. Why things are as they are on the surface of the earth has always intrigued man, but realistic answers awaited the growth of science. Earlier explanations, like the theory that glacial deposits had been rafted into place by icebergs floating in the sea of the Biblical flood, had nonetheless involved careful observation of phenomena. Even if it relied on divine explanations such thinking often showed an acute awareness of interdependence in nature.

The growth of science, together with the urbanisation and industrialisation of increasing proportions of mankind, have seen the birth of *new perspectives* and fields of knowledge. These have been applied to both traditional questions and to new problems created by the changing society within which scientists work.

Geomorphology, (the science of earth surface processes and forms) for example, was for many years concerned with explanation in terms of evolutionary sequences or cycles. This traditional approach was based on a limited understanding of processes. During the twentieth century the scope of geomorphology has suffered contraction and fragmentation. This contraction has been in terms of both time and space, from a concern with broad spans of geologic time to present-day processes and from large-scale landscape evolution to a concern with the detailed workings of smaller components within it, such as slope facets and drainage basins. This *concern with process* has brought with it *changes in methodology*—a growth of experimentation and quantification, for example, which re-integrated geomorphology with the mainstreams of science. Other threads running through more recent geomorphological work have been the growth of systems analysis and a concern with the *relevance* and application of knowledge to the needs of society. How people perceive a landscape, the fact that resources do not merely exist in nature—rather that they are culturally defined—and how resource-management decisions are made, all involve an *interdisciplinary approach*. In looking at such problems some geomorphologists have been brought to the interface between the earth and the social sciences. Finally, there is the general growth of ecological awareness, a realisation of the interdependence of natural and social systems.

Parallel trends in the development of the other branches of physical geography are detectable, and it is hoped that they are reflected in both the content and the approach of this book.

The spirit of the book reflects the kinds of sentiments expressed by Cloud and Gibor—

'What we want to stress is the indivisibility and complexity of environment. For example, the earth's atmosphere is so thoroughly mixed and so rapidly recycled through the biosphere that the next breath you take will contain atoms exhaled by Jesus at Gesthamene and by Adolf Hitler at Munich. It will also contain atoms of radioactive strontium 90 and iodine 131 from atomic explosions and gases from chimneys and exhaust pipes of the world. Present environmental problems stand as a grim monument to the cumulatively adverse effects of actions which in themselves were reasonable enough but that were taken without sufficient thought to their consequences. If we want to ensure that the biosphere continues to exist over the long term each new action must be matched with an effort to forsee its consequences throughout the

ecosystem and to determine how they can be managed favourably or avoided. ... This means that we must continue to probe all aspects of the indivisible global ecosystem.... That is called basic research.'
(*The Biosphere* [W. H. Freeman, 1970] p. 123)

The writer hopes that some of you might be encouraged to pursue your studies further and advance such basic research. For the majority of readers whose interests and careers lie elsewhere, and for whom the detailed factual knowledge is of passing concern only, it is hoped that this book achieves *two aims*. Firstly, that it will act as an introduction to the ways of thinking and patterns of explanation in earth science. Secondly, that it will encourage a questioning approach to the consequences of our actions on the environment. Figure 1 is an example of this approach and the diagram attempts to show the chains of causation contributing to the environmental problems of the Los Angeles basin.

The book is organised into seven chapters, beginning with the lithosphere (the rock envelope of the earth) and the flows and cyclings of energy and water in the atmosphere. Chapter 3 looks at the interaction between atmospheric and lithospheric processes. The

role of running water and the nature of water resources forms the focus of Chapter 4. The operation of the living envelope of the earth—the biosphere—is examined in Chapter 5, and the theme of environmental change over the last 2,000,000 years forms the focus of the penultimate chapter. The book concludes with a brief examination of some of the interactions between man, technology and environment.

An attempt has been made to place illustrations at appropriate places in the text so that they define problems, present evidence and outline the techniques of explanation. Like Fig. 1 many of them complement rather than merely illustrate the written discussion.

Although not intended to be factually comprehensive the book presents a range of problems, explanations and inter-relationships at different scales and in different environments. Finally it should be mentioned that the book neglects statistical techniques, in spite of Dury's fundamental observation that without numerical operations our arguments can sink to the level of medieval disputations! This has been done for two reasons, firstly, the existence of many books on such techniques and secondly, the author's belief that the recognition of themes and questions is of prime importance when beginning a study of earth and environmental science.

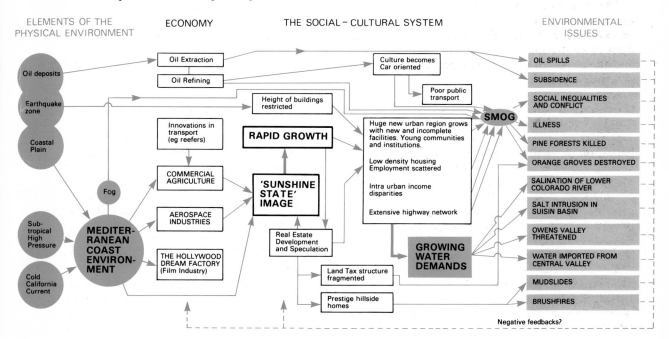

1   Problems of the Los Angeles Basin, interactions of cultural and physical systems

1 The Lithosphere

THE AGE OF THE EARTH AND THE GEOLOGICAL TIME SCALE

If we imagine the earth to be 24 hours old then most of the present oceans are less than an hour old and mankind has been in existence for a mere second. In fact as a planetary body, the earth has existed for 4,600,000,000 years. The nature of the primitive earth is only sketchily known to us and the distribution and character of land and sea have undergone many changes.

If this is the case how has it been possible to unravel the earth's history? Essentially by careful observation of rock sequences, but since rocks are being created and destroyed continuously this is rather like trying to assemble a single complete book from tattered fragments of several copies. To follow this analogy further, rocks do not have page numbers on them so it is the 'story line' which counts. Geologists have therefore used lithological similarities, the fossil record of evolving life and in some cases firm dating (*e.g.* radioisotopes) to assemble the total sequence.

The *geological time scale*, or column, is shown in Fig. 1.1. It is more detailed as the present is approached when the evidence is fresher and more complete. The names applied to the various periods are used from time to time later in the book and the scale should be referred to so that absolute ages and relative positions are known.

THE UNSTABLE CRUST

Earth scientists now conceive of the earth's surface as a series of *plates*, relatively rigid areas of continental and ocean-floor crust. This chapter examines how this view, *Plate Tectonics Theory*, has been formulated and

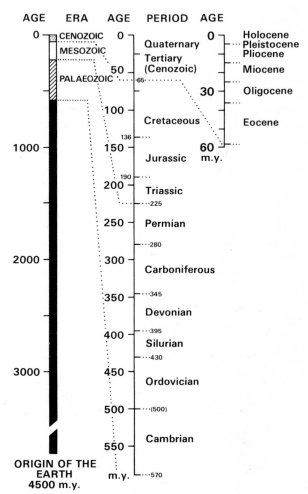

1.1   The geological time scale

accepted. Before looking at how the theory has evolved however we need to look briefly at two phenomena, earthquakes and volcanoes. These have been keys to understanding not only the earth's internal structure, but also the global patterns of fold mountains, sedimentary basins, old stable areas (cratons), marine basins and the 40,000 km system of mid-ocean ridges.

In his search for fuels and minerals man has only superficially scratched the surface of the earth. These activities, together with the examination of rocks exposed at the surface as a result of erosion, have given us knowledge of the uppermost 25 km. Molten rock produced during volcanism (magma) extends our view by giving us insights into conditions up to 100 km beneath the surface.

Earthquakes

'Solid as a rock' is a well-used phrase, particularly by advertising copywriters, but is the earth that solid? Sudden movements of rock occur, in many cases with disastrous effects on man's activities. These earthquakes are a result of the development of stresses and strains which if they cannot be released smoothly result in a sudden break and movement of rock. The location of this fracture is known as the *focus* of the earthquake. The energy released in the break radiates out from the focus in wave form. Figure 1.2 shows these various waves which can be detected on seismographs. Seismology, the study of earthquakes, has given us an insight into the internal structure of the earth. Figure 1.3 illustrates some of this evidence and the interpreta-

tion of the earth's internal layering which is derived from it.

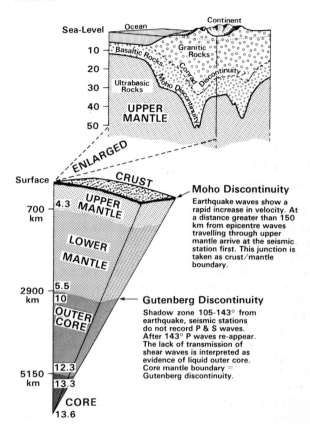

1.3   The internal structure of the earth

Moho Discontinuity
Earthquake waves show a rapid increase in velocity. At a distance greater than 150 km from epicentre waves travelling through upper mantle arrive at the seismic station first. This junction is taken as crust/mantle boundary.

Gutenberg Discontinuity
Shadow zone 105–143° from earthquake, seismic stations do not record P & S waves. After 143° P waves re-appear. The lack of transmission of shear waves is interpreted as evidence of liquid outer core. Core mantle boundary = Gutenberg discontinuity.

Physical geography is concerned with the uppermost layers and the crust has therefore been enlarged in the diagram. The Moho discontinuity, where the waves are accelerated, is taken as the base of the crust. Its depth varies, being shallowest beneath the oceans (typically 5 km), averaging 20 km under continents but increasing to possibly 65 km under large mountain masses. Continental crust is composed of rocks whose density is 2·7 (see Fig. 1.3) whereas ocean-floor crust density is in the region of 3·0. This junction is probably reflected in the Conrad discontinuity whose irregular profile reflects the idea of *isostasy* or balance. A greater thickness of lighter continental crust exerts pressure equal to the thinner and denser oceanic crust. The causes and results of disturbing this state of balance are explored later in the chapter and also

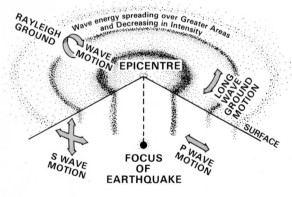

1.2   Earthquake waves

in Chapter 6.

Since earthquake waves travel at different, but known, velocities the distance of the focus from a seismological station can be measured. If records from three stations are available the location of the earthquake can be determined. The number and accuracy of instruments increased rapidly during the 1960's as a result of the attempts to monitor underground nuclear weapon testing. Figure 1.5 shows the distribution of earthquakes. How would you describe this pattern? The key element is recognising seismic and aseismic areas. The quiet, aseismic areas correspond to *plates* and the seismic belts to plate boundaries.

Volcanoes

Volcanism is the term applied to the transfer of molten rock (magma) from within or below the crust to the earth's surface. The variety of volcanic landforms reflects the shifting balance between the constructive processes, magma eruption and the denudational processes attempting to erode them. The actual initial shapes of volcanoes results from an interaction of the mechanism of eruption and the nature of the magma.

This is summarised in Fig. 1.4.

As the magma has been produced in and below the crust the distribution and character of volcanism can give us insights into conditions within the earth, a point which is elaborated in the sections which follow. Figure 1.5 shows the distribution of volcanoes. It is interesting to compare this with the global pattern of seismicity and to reflect how both linear patterns relate to the distribution of the ocean basins, island arcs, continents, mountain ranges and rift valleys.

A volcanic eruption is the most spectacular manifestation of the dynamism of the earth's crust. Events such as the 1883 eruption of Krakatoa in Indonesia (when 25 km³ of the original volcanic island was blown up, with material falling up to 2,500 km away and with the finest dust remaining in the atmosphere for more than two years) are relatively rare. Volcanoes are nonetheless a significant natural hazard in many parts of the world. Taking a longer time perspective they have also been crucial in the evolution of the biosphere, the living envelope around the earth. Without volcanism it is believed there would have been no water and thus no hydrosphere or atmosphere as we know them.

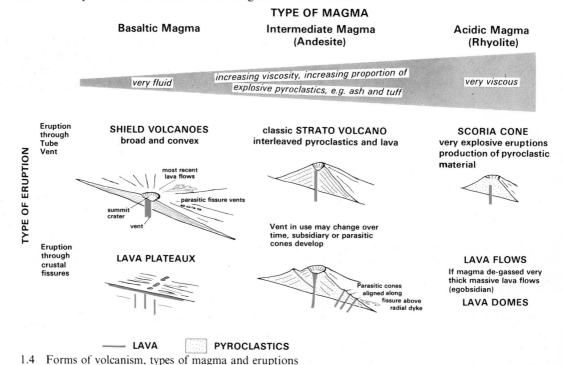

1.4   Forms of volcanism, types of magma and eruptions

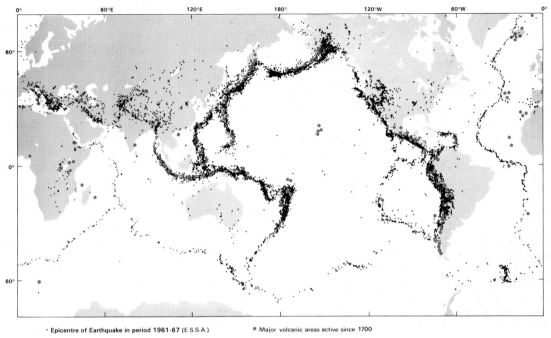

· Epicentre of Earthquake in period 1961-67 (E.S.S.A.)   ● Major volcanic areas active since 1700

1.5   The distribution of earthquakes and volcanoes

PLATE TECTONICS THEORY

Before the emergence of Plate Tectonics Theory there were only 'facts in search of a theory'. These facts, some old and some new, have given us an idea of the forces which have shaped the large-scale features of the earth's surface.

With the development of cartography, from the middle ages onwards the outline of the continents became known and the similarity in shape between the two shores of the Atlantic attracted speculation. As sea-level has changed, present-day shorelines are not the best shapes to compare. The edge of the continental slope, where the sea deepens rapidly beyond the continental shelf, provides a more realistic edge for the continents. An example of this similarity of fit for the southern continents (using the 2,000 m isobath) is shown in Fig. 1.6.

As the study of geology grew, similarities began to be noted in areas now widely separated by oceans, the Great Glen Fault in Scotland and the Cabot Fault in Newfoundland, for example. Parallels in structural trends were also noted between the Appalachians and

Hercynian folded masses of Western Europe (dating from 280,000,000 years ago). In the case of the southern continents an extensive glaciation during Permo-Carboniferous times left the Dwyka series in southern

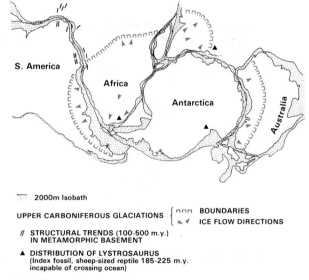

⎽⎼⎽ 2000m Isobath

UPPER CARBONIFEROUS GLACIATIONS ⎰ ⊓⊓⊓ BOUNDARIES
                                ⎱ ▲◄ ICE FLOW DIRECTIONS

// STRUCTURAL TRENDS (100-500 m.y.)
   IN METAMORPHIC BASEMENT

▲ DISTRIBUTION OF LYSTROSAURUS
   (Index fossil, sheep-sized reptile 185-225 m.y.
   incapable of crossing ocean)

1.6   The southern continents

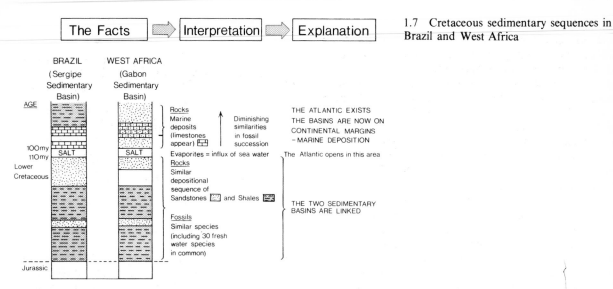

Africa and up to 600 m of similar material in Brazil.

Such facts, and there are many others (see the further readings at the end of the chapter), demanded an explanation. Lower Cretaceous sedimentary similarities in South America and West Africa are shown in Fig. 1.7, together with their later divergence. This type of evidence suggested that both shores of the Atlantic were joined in this area until about 100,000,000 years ago. (Figure 1.7 shows the *evidence–interpretation–explanation* chain.) Imagining alternative explanations throws up problems. In the case of the Lower Cretaceous similarities in Fig. 1.7 if you wish to deny the movement of continents you have to envisage a connecting 'land bridge'. How else would identical freshwater species have occupied both areas? The 'bridge' has to flounder at the end of this period to account for the later divergence. Similar crustal floundering would have to be imagined to explain the origin of the eroded material in the Brazilian glaciation. What this means is that if you wish to deny the ability of continents to move and the relative youth of the Atlantic a range of alternative explanations for such facts has to be presented.

Various strands of evidence like this were synthesised into the theory of *Continental Drift* (notably by Wegener in 1912) which viewed the continents as 'rafts' floating on the denser, yet mobile, lower crust and mantle. This theory failed to be accepted by the 'establishment' within geology—after all the evidence wasn't conclusive and what kind of mechanism could

be envisaged? The idea was kept alive mainly by du Toit in Africa and Holmes in Britain. What caused its rebirth and transformation into Plate Tectonics? In a nutshell: new ideas and new facts.

During the early 1960's, on the basis of some circumstantial evidence and insight, Hess produced the *sea-floor spreading hypothesis*. Hypothesis simply means a statement of expected relationships, only when confirmed by further observation and experiment can it become a theory and used to explain phenomena. Hess synthesised, or pulled together, two facts—that the ocean floor was thin and that a 40,000 km long submarine mountain chain rising 2–3,000 m above its flanking abyssal plain existed. He postulated that the mid-ocean ridge lay astride a rising convection current in the mantle and that new crust was forming continuously with the spreading apart of older crust on both flanks of the ridge crest. Hess also considered when this movement started and suggested that the rate of sea-floor spreading was sufficient to create all the present ocean basins in the last 200,000,000 years.

As the earth can hardly have expanded in volume to allow for the 70% of its surface occupied by oceans to have formed in a mere 5% of its existence, crust must be being destroyed somewhere. Hess suggested that the *ocean trench* systems flanking the Pacific were the sites of descending limbs of convection currents, the whole process of crustal birth and death being similar to a conveyor belt, with birth at the ridge and death in the trench.

Evidence for this hypothesis became available from sources and techniques unknown to the protagonists of the earlier theory of Continental Drift. One of these was *palaeomagnetism*, the study of the earth's magnetic field in the past. It has become apparent since man has used compasses for navigation that the magnetic pole migrates. More exciting, however, the magnetic field itself reverses, having done so at least 171 times in the last 76,000,000 years. Evidence for these reversals became available in the 1960's.

Volcanism was mentioned earlier as a tool to our enquiry. Oceanic volcanism mainly takes the form of undramatic, non-explosive, fissure-type eruptions with the magma cooling rapidly underwater to form so called pillow lavas. The lava involved is known as tholeiite, a basaltic lava believed to form by partial melting of the upper mantle at shallow depths under low pressure and high temperatures. These are precisely the type of conditions one would expect in a rising convective current beneath a ridge crest.

During the cooling of basalts individual minerals sensitive to the magnetic field align themselves to the magnetic pole. Refinements in dating techniques (*e.g.* the potassium argon method) began to allow such basalt flows to be dated. This in turn meant that the dates of magnetic reversals could be fixed. Vine and Matthews exploited this technique in connection with the idea of sea-floor spreading. We can envisage the liquid basalt at the mid-ocean ridge crest being magnetised in the field existing at the time it cooled and solidified. With continued spreading as new crust forms, the basalt with its 'frozen' evidence of polarity moves away on both sides of the ridge. When polarity reverses the basalt now solidifying at the centre of the ridge becomes magnetised in the reversed polarity and in its turn is carried away. This continuous process produces alternate 'stripes' of reversed and normal magnetism in the rocks on either side of the ridge— a kind of tape recording of the earth's magnetic history. Figure 1.8 traces magnetic profiles on either flank of the Pacific ridge crest. They have been drawn in this way so that the symmetry can be seen. Once such magnetic reversals could be dated the *existence* and the *rate of sea-floor spreading* was confirmed.

The world-wide similarity of palaeomagnetic data from either side of the mid-ocean ridges was only one type of new fact which became available. As we know from oil exploitation in the North Sea, drilling tech-

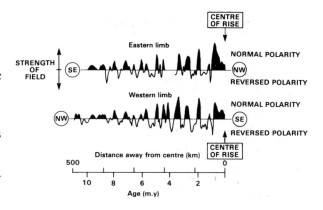

1.8   Magnetic profile across the East Pacific Rise

niques also made advances during the 1960's. Drilling into such shallow waters is, however, considerably easier than drilling into the deep ocean floor, a process likened to drilling through the pavement with a string of spaghetti suspended from the Post Office Tower! Figure 1.9 shows the age of sediments discovered during the JOIDES project. This indicates that the age of the lowest (*i.e.* the oldest in that drill core) sediments increase away from the mid-ocean ridge. Such evidence further confirmed the hypothesis of sea-floor spreading which already seemed likely from palaeomagnetic data.

As you will have noted earlier the mid-ocean ridge is also the site of volcanic and seismic activity. Ridge crests, where new material is being added to the plates, are thus known as *constructive plate margins*. Figure 1.11 shows the structure of Iceland. Notice that the

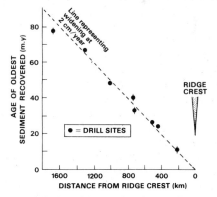

1.9   The age of the oldest Atlantic sea floor sediments recorded on JOIDES leg III (approximately 30°S)

1.10　The eruption of Surtsey, 14th November 1963

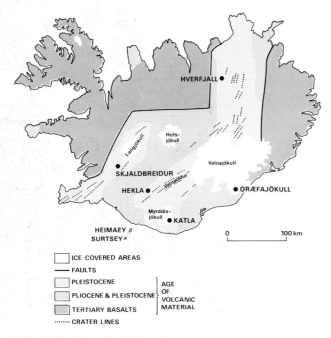

1.11　Iceland: geological structure

youngest material lies in the centre and is flanked by older basalts. If we imagine the two parts of the Atlantic ocean floor moving apart (*i.e.* the American and European plates) then tension will cause cracks parallel to the ridge crest which Iceland straddles. The map shows the location of two such fault systems. Fissures parallel to the ridge axis are sites for volcanism which has an impressive history in Iceland. The Tertiary plateau basalts (of Eocene–Pliocene age) are up to 5 km thick, the number of volcanoes active since the Pleistocene exceeds 200 and the area of lava spreads exceeds 10,000 km². So impressive is this inventory that estimates have been made that since 1500 AD a third of all the lava poured out on the earth has done so in Iceland.

The map (Fig. 1.11) shows the location of some of these eruptions. The Laki fissure eruption of 1783–4 occurring from 100 vents along a 25 km axis poured out 15 km³ of lava. Its associated ash killed 75% of the sheep on Iceland and caused widespread hardship. Some linear forms have been explosive, producing scoria or clinker cones such as the Vatnaöldur chain. Central vents also exist and have produced classic shield volcanoes, *e.g.* Skjaldbreidur and Hverfjall. Composite ash and lava cones are also found such as Hekla and Oraefajökull. Fourteen major eruptions have occurred during this century, two of the more recent being the birth of the island of Surtsey and the eruption of Kirkjufjell on the island of Heimaey in 1973.

It is of course Iceland's position astride the mid-ocean ridge and the constructive plate margin which has resulted in this wealth of evidence for the processes of crustal formation.

Returning now to the broader view, Fig. 1.12 shows the situation in the South Atlantic. It summarises the argument so far and indicates the range of evidence for sea-floor spreading. At this stage it is reasonable to ask just how fast this process is operating. Hess had postulated that the present ocean basins were young compared to the continental portions of plates and the earth as a whole. The oldest sediments recovered from the Bermuda Bank, for example, are 160,000,000 years old—a date in line with his estimates of the age of the oceans. Other evidence indicates rates of 2 cm a year on each flank in the Atlantic and up to 9 cm in the Pacific. Rates fast enough to widen the Atlantic by your height in your lifetime.

In contrast to the constructive plate margin a different type of plate margin is found in the Pacific.

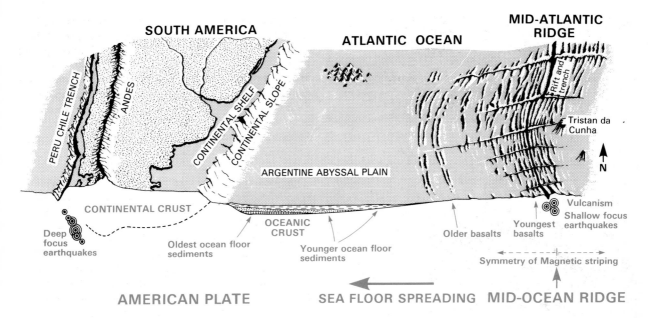

SOUTH AMERICA — ATLANTIC OCEAN — MID-ATLANTIC RIDGE

AMERICAN PLATE — SEA FLOOR SPREADING — MID-OCEAN RIDGE

1.12 The South Atlantic

Figure 1.13 shows the distribution of a range of phenomena in Japan. Eight of the major volcanic eruptions in recent times are shown. Ocean-floor depths are also indicated and these can be related to the ocean trenches, the relief of the Japanese mainland and the arc-like trend of the Ryukyu and Kuril Islands.

Figure 1.13 also shows (in blue) areas affected by frequent and large earthquakes, as well as the location of the seven most destructive earthquakes of the last century. The Kwanto earthquake of 1923, for example, killed 140,000 people and damaged 710,000 homes in Tokyo and Yokohama. The table lists the depths of 21 earthquake foci whose location is shown by the black letters on the map. Can you detect any pattern in the depth of these earthquakes?

Earthquakes reflect the existence of friction and fracturing. This deepening of earthquake foci away from the ocean-floor trench and north-westwards beneath Japan is known as a Benioff Zone. The total picture of seismicity (the depth and distribution of earthquakes), volcanism, crustal movements (shown in black symbols), mountains and sea-floor trenches is interpreted as the destruction of a plate.

The Pacific plate, to the south and east, is being over-ridden and forced down by the leading edge of the Euro-Asian plate. This situation is known as a *de-*

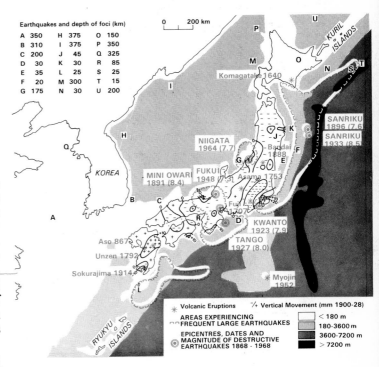

Earthquakes and depth of foci (km)

A	350	H	375	O	150
B	310	I	375	P	350
C	200	J	45	Q	325
D	30	K	30	R	85
E	35	L	25	S	25
F	20	M	300	T	15
G	175	N	30	U	200

1.13 Crustal movements in Japan

structive plate margin. Ocean-floor trenches are the sites where the veneer of ocean-floor crust begins to be carried down the sloping thrust plane of the Benioff Zone. At increasing depths this ocean-floor crust material is fused under conditions of intense pressure and heat and rises as andesite magma. Such volcanism has built classic composite volcanoes such as Mt. Fuji and constructed arcs of volcanic islands. Destructive plate margins are thus 'graveyards' where the oldest ocean-floor crust is consumed in this *subduction zone*. If you look back at Fig. 1.12 its western end portrayed such a destructive plate margin on the other side of the Pacific. Here deep-focus earthquakes indicate the subduction, or drawing down, of crust. Volcanoes have punctured the crumpled continental sediments of the Andes (associated with the westward-moving American plate) and intrusive volcanism has produced massive granitic batholiths in the core of the mountains.

At 5.15 a.m., 18th April 1906, the people of San Francisco were wakened by a major earthquake, 700 of them were killed, many more injured and fires which broke out razed many buildings. A sudden 6·4 m displacement had taken place on the San Andreas Fault which runs under the city. Figure 1.14 shows this fault zone which is an example of the third type of plate boundary, the *transform or conservative plate margin.*

Figure 1.14 indicates that the Pacific plate is moving north-westwards relative to the American plate. In a number of places this movement is achieved by many small adjustments undetectable without seismographs and surveying across the faults. This process is known as creep. But as Figure 1.14 shows, over considerable distances the complex of faults is locked, in which case stress builds up until fracture occurs and an earthquake is recorded. The San Andreas Fault system is creeping overall at a rate of 5–6 cm a year. In areas where the fault is locked, considerable stress has built up—already in the San Francisco area the stress built up is equivalent to that released in the 1906 earthquake. Hopefully earthquake prediction, emergency procedures and new construction standards will result in less loss of life when the locked San Andreas moves again.

Although the fault zone has been described as a transform zone, where one plate is moving past another, the situation is a little more complex than that. The American plate is itself moving westwards. The Sierra and Coast Ranges may therefore represent crumplings on this advancing continental front, whilst volcanism (Shasta, Lassen, Hood and the Crater Lake caldera) reflects friction and melting at the junction of the advancing American plate and the Pacific plate.

Plate Tectonics Theory is in simple terms a synthesis of the kind of ideas presented above. It portrays the forces which have shaped the earth and aids an understanding of its large-scale tectonics. Figure 1.15 shows the seven major plates and several minor plates which make up the surface of the crust. It can be compared with Fig. 1.5. It also shows the location of the mid-ocean ridge systems and classifies the plate margins into the three categories introduced above. Finally it indicates the present direction of plate movement.

The cause of this movement is not yet clear and a range of views exist. In broad terms it may be connected with slow convection currents fuelled by radioactive processes. These could be deep-seated in the mantle or they may be shallow (*i.e.* 100–400 km

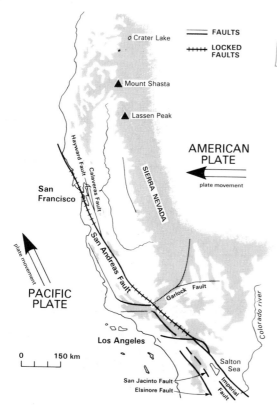

1.14 The San Andreas Fault zone

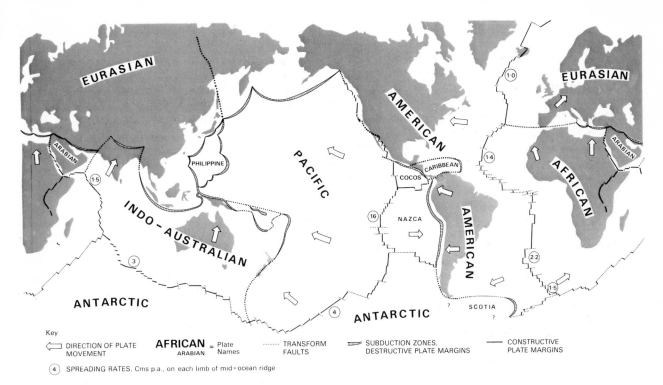

Key

DIRECTION OF PLATE MOVEMENT   **AFRICAN** ARABIAN = Plate Names   .... TRANSFORM FAULTS   ▭▭▭ SUBDUCTION ZONES, DESTRUCTIVE PLATE MARGINS   —— CONSTRUCTIVE PLATE MARGINS

④ SPREADING RATES, Cms p.a., on each limb of mid-ocean ridge

1.15   Plates, plate margins and plate movement

depth). Such currents may be located at constructive margins, pushing up the crust being created which then 'slides' away. Alternatively they may be at destructive margins, pulling crust towards the trenches. In any event they would not appear to be as simple as those postulated by Hess, as is revealed by the existence of offsets and transform faults in the mid-ocean ridge systems. The theory's detailed refinement and an understanding of its causation await further work by earth scientists. It would seem reasonable however to accept the available evidence for the existence of plates and their current movement.

It is fascinating to attempt to reconstruct the *past locations* of the present plates. A different collection of plates may have existed in the earlier periods of earth history. As the earth has been subject to the work of water, waves, wind and ice for a long time the evidence for earlier plate margins is likely to remain conjectural. When we enter the more recent past, however, a clearer picture exists which is of more direct relevance to physical geography.

Figure 1.16A reconstructs the situation 200,000,000 years ago. Notice that what is to become the British Isles lies 10°N of the equator and is experiencing

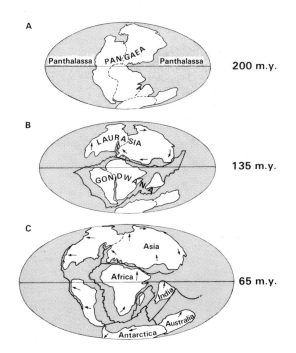

1.16   The history of plate movements

the semi-arid environment of Triassic times. By 135,000,000 years ago the single continent of Pangaea becomes divided into a northern Laurasia and a southern Gondwanaland (Fig. 1.16B). As we have already seen sedimentation and fossil evidence suggest that the Atlantic began to form in lower Cretaceous times (Fig. 1.7). The creation of the Atlantic by sea-floor spreading was well advanced 65,000,000 years ago, as can be seen from Fig. 1.16C.

It is quite revealing to look at 'India's' 8,000 km voyage from the southern polar regions to its present position. What were the consequences of this movement? Figure 1.17 illustrates the structure of India; the peninsular portion has similarities with Australia and Antarctica to which it was originally joined. Moving north 'India' began to impinge on the Euro-Asian plate. Between the plates a marine sedimentary basin was filling with sediments eroded from the adjacent continental surfaces. Such basins of sedimentation are called *geosynclines*. Continued plate movement caused the contortion of these sediments to produce the fold mountain system of the Himalayas.

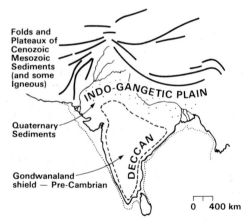

Folds and Plateaux of Cenozoic Mesozoic Sediments (and some Igneous)

INDO-GANGETIC PLAIN

Quaternary Sediments

DECCAN

Gondwanaland shield — Pre-Cambrian

0   400 km

1.17   The structure of India

The creation of *fold-mountain systems* from contorted sediments is not as simple as suggested above. In addition to the surface crumplings there are deeper seated contortions within which the sedimentary rocks are metamorphosed, or changed, by intense pressure and heat (see Chapter 3). Magmatic material may also be injected, cooling to give igneous rocks such as granite. At the surface the mountains are eroded which leaves deep 'roots' of folded lighter continental material embedded in the crust beneath. The weight of material removed by the agents of erosion upsets the isostatic balance and compensating uplift occurs.

The creation of fold mountains in this way, a process known as *orogenesis*, thus consists of two movements—a horizontal folding and contorting force (tectogenesis) and a vertical uplifting force (the orogenic phase). The agencies of weathering and erosion attack these fold mountains, reducing their height and mantling adjacent areas with sediments, as in the case of our example, the Indo-Gangetic Plain.

Plate movement, or—to put it more dramatically—plate collision, can therefore be used to explain the character and distribution of fold-mountain systems. Sediments accumulate in geosynclines, which have often continually deepened, as is indicated by extensive thicknesses of shallow water facies. During the tectogenesis stage the sediments are contorted into simple folds or with increased severity of movement into recumbent folds. Nappes develop when fracture and slippage occurs along the axis of such contorted folds. The whole process is associated with volcanism as, during the deformation and heating associated with mountain building, some of the geosynclinal sediments melt. They may rise to crystallise as granite *batholiths*, which may later become exposed at the surface following the erosion of overlying rocks and the processes of vertical uplift. These kinds of processes have and are operating at the plate boundaries discussed earlier (Figs. 1.12, 1.13 and 1.14) but there are detailed differences which are beyond the scope of this general discussion.

Figure 1.18 shows the pattern of the major *physiographic divisions* of the earth which can be interpreted in terms of the processes underlying plate tectonics. Compare this map with Fig. 1.15, which showed the location of the plates and their current movement, and with Fig. 1.16. It is important to remember that plates are not merely the continents, they consist of areas of oceanic and continental crust. Earlier in the chapter case studies of the three types of plate margin were presented. With this information and the discussion of mountain building in the section above it should now be possible to relate plate patterns and dynamics to the physical forms of the oceanic and continental surface.

Bearing in mind the span of geological time in relation to our lifetimes it is obviously not relevant to speculate too deeply on the *future plate positions*.

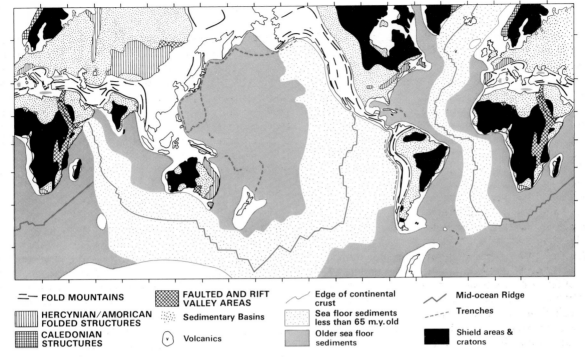

☰ FOLD MOUNTAINS	⊞ FAULTED AND RIFT VALLEY AREAS	Edge of continental crust	Mid-ocean Ridge
▥ HERCYNIAN/AMERICAN FOLDED STRUCTURES	⸬ Sedimentary Basins	Sea floor sediments less than 65 m.y. old	Trenches
⊞ CALEDONIAN STRUCTURES	ⓥ Volcanics	Older sea floor sediments	■ Shield areas & cratons

1.18   World structure

Nonetheless it is quite stimulating. Figure 1.19 is a projection for 50,000,000 years ahead. The Atlantic widens, Africa continues its northward migration and Australia straddles the equator, which might ease its water supply problems but might play havoc with its cricketing weather. Finally if the San Andreas Fault continues to behave as it does today the luckless inhabitants of an earthquake-proof Los Angeles would find themselves in the latitude of Southern Alaska and on their way to ultimate extinction in the Aleutian Trench!

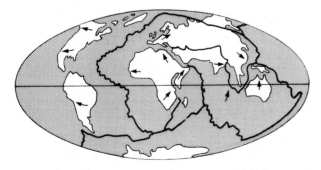

1.19   The possible positions of plates in 50,000,000 years

Review Questions

1. Evaluate the evidence for Plate Tectonics Theory.
2. Describe and explain the distributions of earthquakes, volcanoes and fold mountains.
3. The discussion in this chapter has ignored the Indian Ocean. If you had to plan an investigation by earth scientists to confirm the ideas of plate tectonics, where and what would you attempt to investigate?
4. Using other references investigate the character, distribution and origins of rift valleys and shield areas (cratons).

Further Reading

Bolt, B. A., 1978, *Earthquakes—A Primer*, W. H. Freeman.

Dineley, D., 1975, *The Earth's Voyage Through Time*, Paladin.

Elder, J., 1976, *The Bowels of the Earth*, Oxford University Press.

Holmes, A., 1978, *Principles of Physical Geology*, Nelson.

Wilson, J. T., 1977, *Continents Adrift and Continents Aground: Readings from Scientific American*. W. H. Freeman.

2 The Atmosphere

INTRODUCTION

We all have plenty of evidence of how the atmosphere impinges on man. It might be mundane television news scenes of traffic jams caused by people fleeing the hot and gritty feel of summer in the city. It could be a rise in food prices reflecting reduced supplies because of unusual weather. At a more serious level it could be news of death and destruction caused by hazards such as cyclones and floods.

The need to know about atmospheric processes is therefore obvious. With 70% of the earth covered in ocean and with large areas of land sparsely populated remote environmental sensing from satellites has given scientists an almost instantaneous picture of global weather for the first time. Such a massive explosion of data does not, however, wholly explain our increased knowledge of the atmosphere. Of equal importance is our ability to process, understand and act on the information obtained. The production of *models* and the use of computers have thus been equally vital tools for understanding the behaviour of the atmosphere.

The word model may need some explanation. A model can be a diagram, an analogue, a piece of hardware or an equation. All are essentially simplifications of reality which help us understand how things work. In the case of the atmosphere mathematical models are most powerful, particularly as they allow prediction and forecasting. With the sheer size of the atmosphere and the number of variables to be considered, it is not surprising that only computers can model the atmosphere quickly and also at global scale. Such models are beyond the scope of this book. Some simple descriptive models will be introduced in the sections which follow.

Man's knowledge of the atmosphere might have increased but his ability to act on the basis of this knowledge is all too often impaired. Firstly he may not be rational. His perception of the size and frequency of a hazard, such as drought or storm damage, may not match reality. Secondly knowledge may exist but a communications breakdown may not allow it to reach the people or area affected. This failure of communications could be technical or it could reflect the social structure and the way information flows through the community. Weather satellite ITOS 1 in November 1970 detected a cyclone in the Bay of Bengal. The effect of its high winds driving 10 m high waves in a storm surge onto the flat land of the Ganges Delta could be predicted. Sadly this knowledge couldn't be communicated to the rural areas in time and almost 500,000 people died.

Finally man must not be viewed as a passive recipient of atmospheric hazards and bonuses. He has increasingly modified the atmosphere, indirectly by burning fossil fuels and changing its composition and by clearing natural vegetation. His attempts at direct modification through such 'weather engineering' as cloud-seeding have been on a much smaller scale.

The character of the atmosphere

The atmosphere is the envelope of gases held to the earth by the force of gravity. By volume it is composed of nitrogen 78%, oxygen 20.9% and argon 0.9% together with small and variable amounts of carbon dioxide and water vapour. As we will discover later the concentration of the latter two gases is of great significance to mankind. Solid particles from such sources as volcanic eruptions, sea salt and even blown soil are also a small but vital constituent of the atmosphere.

Although the outer edge of the atmosphere may be taken as being 10,000 km above the surface, almost 97% of it lies within 29 km of the earth. Figure 2.1 indicates the structure of the atmosphere, the names applied to the various zones and the changes of temperature with height. The two lowest layers, the troposphere and stratosphere are our particular concern in this chapter. The boundary between them (the tropopause) can be found on the temperature height diagram at the point where the fall of temperature with height detectable in the troposphere, changes to the isothermal conditions of the lower stratosphere, (*i.e.* where temperature is unchanged with height). The height of the tropopause varies from approximately 8 km at the poles to 16 km at the equator.

ENERGY IN THE ATMOSPHERE

The sun

The sun is our nearest star, 150,000,000 km from the earth. All weather, winds, clouds and rain are the consequence of the sun's emission of energy, the twelve-month circulation of earth around the sun and the daily rotation of the earth on its inclined axis.

The sun is the prime energy source for the earth. Energy can be defined as a body's capacity to do work, it exists in various forms, three of which concern us in this chapter—radiant, thermal and kinetic. Within the sun, thermonuclear processes convert hydrogen to helium. From the surface layer of the sun, which has a temperature of 6,000°C, energy is transmitted in all directions as radiant energy or *radiation*.

Electromagnetic radiation can be described in terms of its wavelength and its frequency. The sun's radiation covers a wide spectrum ranging from the short wavelength and high frequency gamma and X rays, through the ultraviolet and into the visible spectrum. At longer wavelengths are the infra-red and radio waves. Figure 2.2 indicates the wavelengths of solar radiation and the rate of energy emission. As you will see from the shape of the emission curve, the sun's energy is not evenly distributed, approximately 45% occurring in the visible spectrum.

Radiant energy from the sun travels through space at the speed of light until it is intercepted by a gas (*i.e.* the atmosphere), a liquid (water) or a solid (*i.e.* the earth's surface or particles in the atmosphere). When intercepted it is either absorbed (changing from radiant to thermal energy) transmitted or merely reflected unchanged back to space. In the section which follows the flow of this energy will be traced.

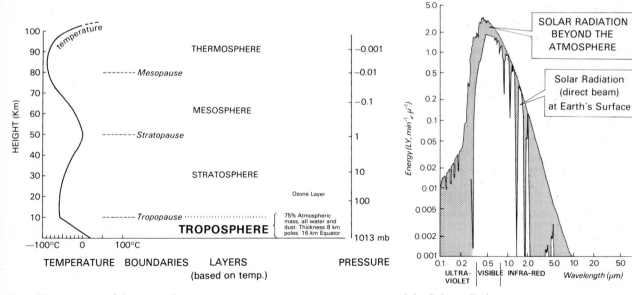

2.1   The structure of the atmosphere

2.2   Solar radiation

The solar energy cascade

Energy from the sun is the prime source of energy for the earth. If we initially imagine an earth without an atmosphere the amount of solar energy received at the surface will depend upon four factors—the solar output, the distance from the sun, the apparent altitude of the sun in the sky and the length of day.

Firstly the solar output. The earth intercepts only a fraction of the total radiation emitted in all directions from the sun. The rate of this emission reaching the earth is known as the solar constant and is approximately 1.96 cal/cm²/minute. Although the word constant is used there are variations of up to 2% in the sunspot cycle. Apparently unimportant in influencing daily and yearly weather, this variation may be considered when long-term global climatic change is discussed in Chapter 6.

The second factor of distance is related to the fact that the earth's orbit is eccentric, being closest to the sun on 3rd January (the perihelion) and farthest on 4th July (the aphelion). This changing distance produces a variation from 1.88 cal/cm²/minute at the aphelion to 2.01 cal/cm²/minute at the perihelion. The effects of this are, however, masked by the movements of the atmosphere and oceans.

The third factor is the 'altitude' of the sun, a result of the spherical shape of the earth. A bundle of solar radiation approaching the earth at 60°N, for example, affects a surface area twice as large as it would at the equator.

Finally, of course, the earth's axis of rotation is inclined $23\frac{1}{2}°$. This means that poleward of $66\frac{1}{2}°$ (the Arctic and Antarctic circles) part of the year has no solar radiation. Between $66\frac{1}{2}°$ and $23\frac{1}{2}°$ (the Tropics) radiation curves show one maxima and one minima, whilst equatorwards of $23\frac{1}{2}°$ the sun appears to pass overhead twice a year and the radiation curves show two maxima and two minima. Figure 2.3 portrays the effects of these factors on the radiation receipts from month to month at various latitudes.

In the above discussion we simplified conditions by assuming that the earth had no atmosphere. What happens when we trace the cascade of solar energy through the atmosphere towards the surface?

Firstly at elevations of between 20–35 km the solar radiation passes through the *ozone layer*, where photochemical processes have resulted in a region of concentration of oxygen molecules O_3. In this zone ultra-

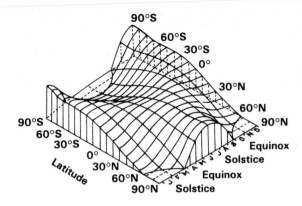

2.3   Global insolation variations with latitude and season (assuming no atmosphere)

violet radiation is absorbed. This absorption produces heat (as can be seen from the temperature height data in Fig. 2.1) which provides energy for the upper-air circulation. The ozone layer also acts as an important 'umbrella' shielding life at the surface from damaging ultraviolet radiation.

Figure 2.4 attempts to show what happens to the incoming solar radiation when it enters the troposphere. Carbon dioxide and water *absorb* a portion, especially the infra-red radiation, resulting in a rise in the temperature of the air. Some of the radiation is reflected from clouds or the ground surface. The proportion of radiant energy reflected back by the surface is known as the *albedo*. The albedo of cloud tops ranges between 30–80%. The earth's reflectivity itself is highly variable ranging from 9% over coniferous forest to over 35% above sandy desert areas. The sea's albedo is even more variable, being low when the sun is high and the sea rough, and very high with a low-angle sun and a calm sea. Satellite observations show an annual average albedo for the earth as a whole of between 27–34%, with surprisingly little variation between hemispheres in spite of the large differences in the relative proportions of land and sea. This points to the importance of cloud cover in determining the global albedo.

On its passage through the atmosphere the solar radiation is *scattered*, *i.e.* the rays are turned aside in all directions as they collide with molecules of atmospheric gas. Since scattering occurs in all directions, a proportion reaches the earth's surface as downscatter or diffuse sky radiation, while the remainder is scattered back into space. Scattering effects are greatest at the shorter wavelengths of the visible spectrum; thus we see this diffuse sky radiation as blue sky.

Figure 2.4 shows that after scattering, reflection and absorption 24% of the incoming short-wave radiation reaches the surface directly. An additional 21% penetrates as diffuse sky radiation, so that the earth's surface receives 45% of the incoming radiation which was available at the top of the atmosphere. Only 24% is actually absorbed within the atmosphere itself.

Terrestrial radiation and heat transfers from earth to atmosphere

The incoming solar radiation, both direct and diffuse, which reaches the surface is absorbed and raises the temperature of the earth. The surface in turn radiates this energy back to space. The earth's surface is cool compared to the sun, averaging 27°C, and the terrestrial radiation is thus long-wave with wavelengths between 3 and 50μ. This outgoing terrestrial radiation continues during both day and night.

In fact it is a little too simple to view the earth as giving energy to the atmosphere by long-wave radiation alone. There are two other mechanisms by which heat is transferred to the atmosphere—latent heat and direct conduction. The *latent heat transfer* is linked with the evaporation of water, during which process energy is stored to be later released when condensation occurs. The release of heat when condensation occurs raises the temperature of the air. Water vapour moves horizontally and vertically in the lower atmosphere and this movement is thus an important part of the heat transfer processes at work between the earth and its atmosphere. Direct *conduction* occurs when heat is transferred from a land or sea surface to the air in contact with it. This process is aided by turbulence in the lower atmosphere.

To return to the long-wave outgoing terrestrial radiation. In an analogous manner to the selective absorption of certain parts of the short-wave incoming radiation, so too, various wavelengths of the outgoing radiation are absorbed. Carbon dioxide and water vapour in the atmosphere absorb radiation between the 5–8μ and 12–20μ wavelengths. Such absorption raises the temperature of the lower atmosphere which in turn radiates energy in all directions, some being radiated back to earth, some to adjacent zones in the

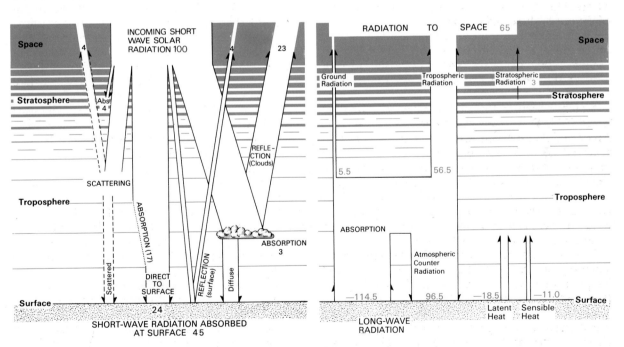

2.4 The solar energy cascade

2.5 Terrestrial radiation and heat transfers

atmosphere and eventually, as atmospheric density decreases with height, some is finally lost to space. To add to this loss of course we have the unabsorbed radiation from the surface (between 8 and 11 μ wavelength) which passes directly to space.

The absorption of some of the outgoing long-wave radiation by carbon dioxide and water in the atmosphere is known as the *'greenhouse effect'*. The incoming short-wave radiation passes through the atmosphere while much of the outgoing long-wave radiation is absorbed and energy effectively trapped. Figure 2.5 summarises the processes of heat transfer from earth to atmosphere which have been discussed above.

The heat budget of the atmosphere

So far we have considered the incoming and outgoing radiation separately. We will now examine their combined effects. The earth's geological record (*i.e.* the history of the evolution of life), shows that the earth is neither cooling nor warming to any appreciable extent. The radiation received from the sun over the long term must therefore be balanced by the outgoing losses from the earth.

Figure 2.6 indicates the *net radiation* at different latitudes. If we look at the figures for the earth's surface (the broken line) they indicate a net radiation surplus. In other words the incoming radiation from the sun, both direct and diffuse, and the long-wave downward atmospheric counter-radiation exceeds the outgoing radiation, comprised of reflected short-wave and long-wave radiation from the surface.

For the atmosphere alone on the other hand (the dotted line) there is a net deficit. If we combine these two to represent the combined earth–atmosphere system (the solid line) you will notice the surplus in low latitudes and the deficit in high latitudes.

The effects of the earth and atmosphere energy budgets are twofold. Firstly the deficit poleward of 40° for the combined earth–atmosphere system results in a poleward heat transfer, the *meridional heat flux*, accomplished by movements of the atmosphere and oceans. Without this meridional heat flux the poles would be getting colder, the equatorial regions hotter! Secondly, as the earth's surface is not warming and the atmosphere cooling, there is a heat flow from the earth's surface to the atmosphere. This *vertical heat flux*, as was mentioned earlier, is

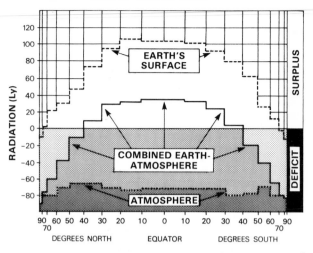

2.6 Pole to pole profile of net all wave radiation for A. earth's surface, B. atmosphere and C. combined earth-atmosphere

accomplished by radiation, conduction and latent heat transfer.

There are of course variations in detail imposed upon this generalised picture. The effect of diurnal (day–night), and summer–winter changes in the radiation picture are shown in Fig. 2.7. In the case of this example the June noon figure is four times greater than December's and the period of radiation is ten hours longer. The effects of such variations in intensity and duration of radiation on temperature can be seen in the daily and monthly temperatures of middle and high latitudes.

Figure 2.8 indicates the pattern of *mean annual solar radiation* received at the ground surface. It reflects the operation of a number of controls. The simple latitudinal effect is distorted in parts for example by cloudiness and the thickness of the atmosphere. In connection with the latter point the thickness of the troposphere varies from 8 to 16 km from pole to equator. What would be the effects of this on the solar energy cascade in various regions?

At a much smaller scale there are variations in solar energy receipt according to the aspect of a slope and its angle. Figure 2.9 represents the direct solar radiation receipt on slopes of northerly, southerly and easterly aspects. Notice the considerable variations between slopes of different inclinations but with the same aspect, or different aspects but with the same

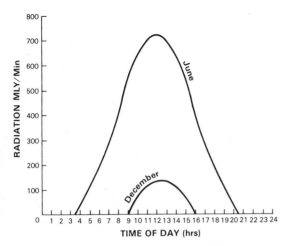

2.7   Solar radiation curves for Hamburg, 54°N

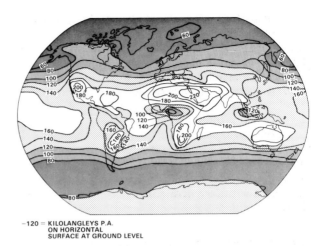

−120 = KILOLANGLEYS P.A.
ON HORIZONTAL
SURFACE AT GROUND LEVEL

2.8   Global mean annual solar radiation

degree of slope. In hilly terrain such effects are often noticeable in the vegetation cover. Figure 2.10 shows the contrast in forest areas on opposite sides of the east–west trending Otta Valley. Cultivation here at 62°N has only been attempted on the south-facing

slopes. Such *microclimatic* effects are significant at the margins of cultivation, the vineyards of the Rhine valley, for example, have been 'microclimatically transported' several hundred kilometres south by their southerly aspect.

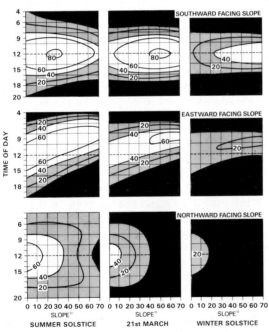

2.9   Slope aspect, angle and radiation. Direct solar radiation (cal/cm²/min) on cloudless days on north, east and south-facing slopes of various inclinations at 50°N shown for three days

2.10   The Otta valley, Norway, with cultivation on the south-facing slope

MOISTURE IN THE ATMOSPHERE

Water is a vital constituent of the atmosphere. It is essential for life and in vapour form it also plays a significant part in absorbing radiation and in transporting energy from the surface into the atmosphere.

The global balance of water and its cycling

Figure 2.11 indicates the global balance of water. The oceans covering 71% of the earth's surface contain 97% of all water, of the remainder (the fresh water) the ice sheets are the dominant element, large enough in fact to feed all the world's rivers for 900 years! As R. L. Nace wrote:

'With 97% of the world's water in the sea and 2% in deep freeze, the world evidently is a fine place for whales and penguins, but it has its shortcomings for man. In addition, 17% of the land area is under ice or frozen, and 32% is arid or semi-arid. Small wonder that man, throughout his history, has sought ways to interfere with the water cycle.'

(*Water, Earth and Man*, p. 40)

Man's interference will be dealt with in later sections of the book. The global water cycle itself is shown in Fig. 2.12 which indicates the transfers of water between the stores depicted in Fig. 2.11. This is a global picture, the figures indicating proportions of the world mean annual situation. There are considerable variations both in time and in space, as will be discovered later. It is useful nonetheless to imagine the 'unseen rivers aloft' equivalent to their surface cousins

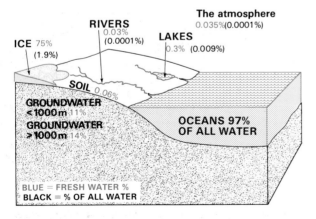

2.11   Global water stores

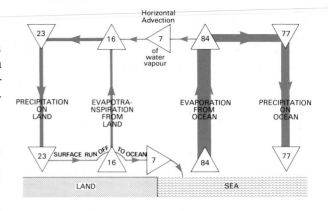

2.12   The global hydrological cycle

maintaining the cycling of water.

Humidity

The concentration of water vapour in the atmosphere is known as humidity. When the water-vapour content is at its maximum the air is said to be saturated. This point of saturation is controlled by temperature. Figure 2.13 indicates how it varies. Notice that cold air, when saturated, holds quite small amounts of water compared to warm air (*i.e.* 4·4 gms/m³ at $-1°C$ and 25 gms/m³ at 27°C). This of course means that cold air is always dry in absolute terms.

Another way of looking at this is to consider that part of the total barometric pressure which is due to water vapour alone. The mass of the atmosphere above a point on the surface exerts a pressure which is measured in millibars. The vapour pressure is the term used to define that part of the total barometric pressure resulting from water vapour alone. This ranges from 2 mb over Siberia in winter to in excess of 30 mb over the Tropic of Cancer in July. It varies obviously with the presence of water sources. These include evaporation from sea, lake and river surfaces; evaporation from ice surfaces; and vapour which has passed from the soil through the roots and leaves of plants to the atmosphere in the process of evapotranspiration. Humidity will also vary with the extent and magnitude of any horizontal transfer of moisture into the region and finally of course, as indicated in Fig. 2.13, with the temperature of the air.

The ratio of vapour pressure to saturation vapour pressure expressed as a percentage is known as the *relative humidity* of the air. As a measure of the degree of saturation it is a useful figure but, as you will realise from what has been said above, without an indication of the temperature it tells us nothing about how much water vapour is actually present in the air. A final term to introduce at this stage is the dew-point temperature, this is the temperature at which air would become saturated (*i.e.* 100% RH) assuming it was cooled at a constant pressure and with no addition of moisture.

Figure 2.14 shows the mean atmospheric water-vapour content in January and July. The measure plotted is mm of water. The global average is 25 mm. What do you notice about the areas above and below this mean in the two months? This water content of the atmosphere is equivalent to approximately ten days' supply of precipitation, a figure which indicates how frequent and intensive the cycling of water shown in Fig. 2.12 must be.

Condensation

In the section above you were introduced to some terms used in connection with atmospheric humidity, to its sources and global supply pattern. It is when it is in its condensed liquid form, however, that atmospheric water has its most direct influence on man.

Surprisingly, in view of what has been mentioned above, air can be supersaturated, continuing to hold water in vapour form after the dew-point has been passed. To maintain a water droplet $0.001\ \mu$ in radius, for example, requires in pure air an RH of 320%. The atmosphere, however, is not a clean laboratory and the air is not merely a mixture of gases! A multitude of *condensation nuclei* exist in the atmosphere. These are hygroscopic particles (wettable substances) which attract water-vapour molecules when the moisture content nears saturation point. Dust, sea salt from evaporated spray, sulphuric acid from combustion and volcanism are such hygroscopic particles. Although invisible to the naked eye they are very numerous, *e.g.* 10^6 per cc in urban areas.

Condensation from vapour to water droplet around the condensation nuclei occurs as the air is cooled. This can be achieved in several ways. Under some conditions cooling can be achieved by *radiation*. If we imagine a night with clear skies, the net loss of radiation from the ground results in a fall in its temperature. Air close to the ground surface is cooled by conduction and radiation of heat from the warmer air to the coo-

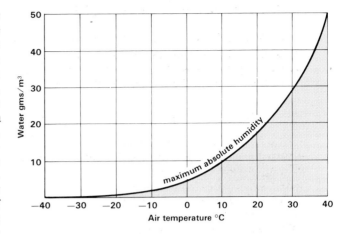

2.13   Air temperature and absolute humidity

ling ground. The air thus becomes cooler and as it approaches its saturation point water droplets begin to form around the nuclei. This type of cooling, with clear skies and light winds to help mix the cooling lower layers of air, forms a layer of water droplets near ground level resulting in radiation fog or mist.

Far more widespread methods of cooling are those connected with upslope motion and convection. The simplest form of *upslope motion* occurs when an air stream meets a mountain barrier and the air is forced to rise. As altitude increases the lowering of pressure allows the air parcel to expand. Individual molecules of gas are more widely spaced, not striking each other

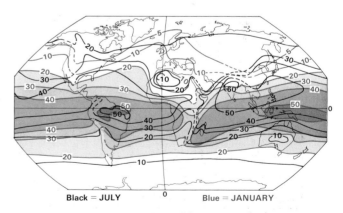

2.14   Water in the atmosphere. Mean atmospheric water content in mm of precipitable water

as often, and this imparts a lower sensible temperature to the air. In the rising and cooling air humidity increases, dew-point is approached and condensation begins, resulting in a cloud mantling the upper slopes. A similar form of upslope motion occurs at a front, a sloping surface separating two air masses. Air masses are bodies of air which have similar temperatures and humidities horizontally over extensive areas. If we imagine a warmer and less dense air mass adjoining a colder and denser air mass the warm air tends to glide upwards over the colder air. Motion up such sloping frontal surfaces results in large-scale condensation.

Convective cooling occurs with the rise of thermals. Solar radiation pouring onto a varied surface will cause differential heating. Air in contact with warmer patches will 'bubble' upwards. The ascending expanding and cooling bubble of air in the thermal may cool sufficiently for condensation to occur and a cloud to form. Glider pilots trying to gain altitude will search for such thermals. Under what sort of conditions and over what kind of ground surface would you expect to find them?

Finally there are two other processes which result in condensation. *Advection* cooling occurs when there is a horizontal transfer of warm moist air over an adjacent cold surface. In this case direct contact cooling occurs and condensation may result in the lower air layers. As with radiation fog, this is of restricted occurrence and limited vertical extent. Advection fogs occur where air above a warm current may cross to lie above a cooler current. In the case of Newfoundland, the Gulf Stream at 14°C and the Labrador current at −1°C provides an ideal setting for the formation of such fogs. Finally in some situations *mixing* of air may occur. At a frontal boundary zone, for example, the characteristics of the two parent air masses can be combined in such a way that condensation occurs.

Stability and instability

We normally find that as we rise through the troposphere temperature falls. This change is called the *environmental lapse rate* (ELR), which is best viewed as the still air lapse rate (*i.e.* the change in temperature which you would experience in a balloon ascent). The ELR varies from place to place and from time to time.

In the discussion above, of convective uplift in a thermal and its associated cooling and condensation, mention was made of the fall in temperature of the rising air. This fall in temperature is known as an *adiabatic lapse rate*, adiabatic indicating that the reduced pressure is the cause of temperature change not the loss of any heat to the surrounding air. In fact there are two adiabatic lapse rates. If the rising air parcel is unsaturated it cools at the *dry adiabatic lapse rate* (DALR), which unlike the ELR is a fixed rate of 9·8°C per 1,000 m. In many situations of course the rising and cooling air may approach its dew-point temperature and condensation begins. If the air continues to rise it will also continue to cool—but at a lower rate, resulting from the release of latent heat as vapour is condensed into droplets. The rate at which this air cools is known as the *saturated adiabatic lapse rate* (SALR), which ranges between 4 and 9°C per 1,000 m at high and low temperatures respectively.

We can represent these three lapse rates on temperature height diagrams. Figure 2.15A indicates a position of *stability*. Notice that the ELR lies to the right of the DALR. If a parcel of unsaturated air is forced to rise it cools at the DALR (*i.e.* 9·8°C/1,000 m). The temperature of the surrounding air is indicated by the ELR line. In the case of our example, at a height of 500 m the rising air parcel will have cooled to 5·1°C but it will be surrounded by air at 8°C. Therefore if the uplifting force is removed the parcel of air, being colder and denser than its surrounding air, will sink back to its starting point.

In the case of our example Fig. 2.15B indicates that at a height of 800 m this rising air reaches its dew-point, above which of course it cools at the SALR (which for illustration has been plotted as 6°C per 1,000 m). Although the SALR is slower than the DALR at an altitude of 1,100 m our rising air parcel is still cooler than the surrounding air. Without its uplifting force it would thus still sink back. Figure 2.15B therefore also represents a condition of stability.

The ELR however is variable. Figure 2.15C indicates the situation with a different ELR (*i.e.* a different time or a different air mass). Notice that the rising air parcel initially cools at the DALR, reaching its dew-point at 6°C. At this height the ascending air is cooler than its surroundings. However, above the condensation level, as it continues its rise, it cools at the SALR and you will notice from the diagram that its temperature progressively approaches that of the surrounding still air. At an elevation of 1,200 m it reaches the same

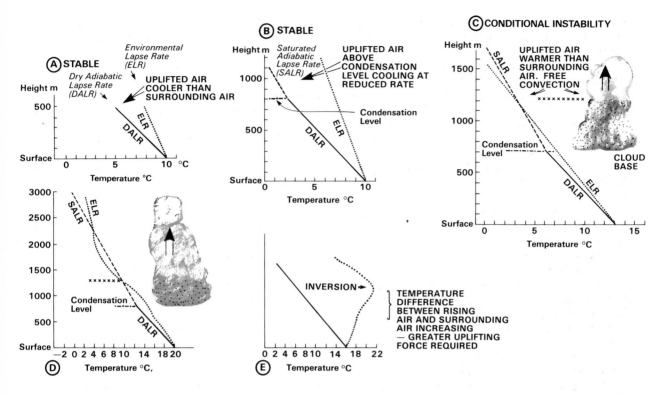

2.15   Lapse rates, stability and instability

temperature as the surrounding air and above this height it is warmer and less dense. It will thus rise freely, even if the uplifting force is removed, and is said to be *unstable*. In this example where the air becomes unstable its buoyant freedom is conditional upon its saturation, this is known as conditional instability. Such an air mass whose ELR is intermediate between the SALR and DALR frequently gives cumulus cloud and showers if it is disturbed.

Figure 2.15D indicates a more realistic situation where the ELR, until now simplified as being a constant rate, varies with height. Notice again the cloud base is controlled by the condensation level, where dew-point is reached. The upper surface of the cloud may lie at the top of the zone of instability, although buoyancy in the rising thermal can of course extend this upwards.

In most cases there is a fall of temperature with height. There are occasions however when a temperature *inversion* exists and there is a rise of temperature

with height, shown in Fig. 2.15E. If you look at the lapse rate curves you can see the inversion acting as a lid or barrier to uplift.

Condensation forms: clouds

Clouds are composed of water droplets, ice crystals or indeed a mixture of both, depending upon the temperatures prevailing aloft. Water droplets characteristically give sharp and distinct boundaries to relatively dense clouds. Ice crystals, on the other hand, give a fibrous and filmy appearance. Although clouds are almost infinitely variable, they are conventionally *classified* on the basis of their shape and structure. They may, for example, be layered (or stratiform) or globular (cumiliform); they may be thin or have extensive vertical development and finally their bases may occur at different heights above the ground. Figure 2.16 shows the basis of cloud classification together with their names and conventional abbreviations.

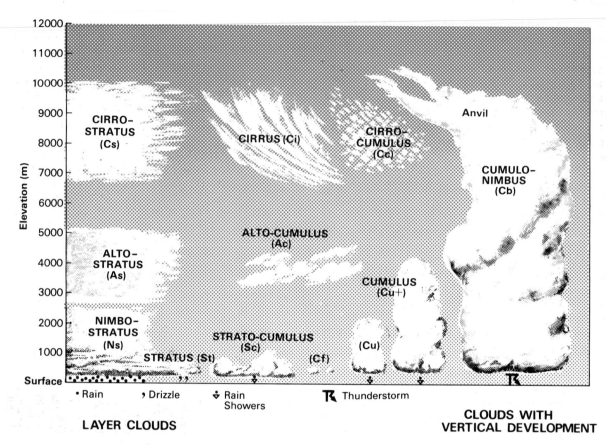

2.16   Cloud classification by height, vertical development and form

Precipitation

In the sections above we examined the processes of atmospheric cooling and the growth of water droplets to form clouds. It is a misconception that condensation is equivalent to precipitation, after all not all clouds produce rain. Cloud droplets are small, $1–50\,\mu$ in radius and held aloft by air movements. The average raindrop is a million times bigger, $1,000–2,000\,\mu$ radius. Just how the cloud droplet is transformed into the raindrop is therefore a fundamental question.

Taking up the misconception mentioned above we might initially assume that the droplet merely continues to grow through further condensation. Figure 2.17 shows droplet growth through time. How long would it take for a cloud droplet to grow into a raindrop? Clouds form and raindrops fall in much *less*

time than this. Something must be causing the *fusion*, or joining together, of the droplets.

One explanation of this fusion is the *Bergeron Theory* which assumes that water droplets and ice

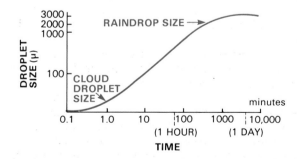

2.17   Water droplet growth by condensation

crystals co-exist within a cloud. At −10°C only one droplet in a million is in fact frozen, at −40°C virtually all droplets are frozen. Between these temperatures the cloud contains a mixture of ice crystals and supercooled water droplets. The theory next notes that the relative humidity of air is greater with respect to an ice surface than a water surface. What this means is simply that the ice crystals can grow rapidly at the expense of the supercooled water droplets. The growing ice crystals then fracture in the moving air currents within the cloud and each fractured limb in turn provides an additional focus for crystal growth. The aggregation of crystals continues and snowflakes form, which become so dense that they fall through the cloud. In falling they pass through warmer air, melt and reach the ground as raindrops.

There are a number of lines of evidence for this theory. In temperate latitudes appreciable precipitation falls only from clouds which extend above freezing level. Lower clouds, such as stratus (see Fig. 2.16), produce light drizzle whilst cumulo-nimbus clouds produce heavy rain. Secondly, radar observations indicate a vast increase in snow in the sectors of clouds where temperatures lie between 0° and −5°C. The partial success of rainmaking provides further confirmation. Clouds rising above the freezing level (containing supercooled water droplets) when seeded with dry ice or silver iodide have often produced rain. The dry ice (frozen CO_2) and silver iodide have acted as substitute ice crystals and 'fooled' the droplets.

Attractive as this theory is in the case of middle latitudes with mixed clouds (*i.e.* ice and water) alternative explanations are needed for tropical areas, where clouds may not penetrate the freezing level yet nonetheless produce copious rainfall. In such *warm clouds* 'super' condensation nuclei may exist, *i.e.* sea salt particles, which retain their hygroscopic attraction to vapour even as the droplet grows. Differential electrical charges may also promote rapid droplet growth.

Once formed the larger and heavier droplets begin to fall. As they do they collide with smaller droplets and also sweep them into their wake, thus further increasing in size. They may become so big that they become unstable and break up so that the whole process is repeated with their offspring. This *collision mechanism* also operates with droplets produced by the Bergeron process but logically can only be significant if the cloud is deep, exceeding 2,000 m in thickness.

There are various *forms of precipitation*. Snow consists of branched hexagonal crystals which have become matted into flakes. Under very cold conditions the crystals lack the veneer of supercooled water to cement the flake and thus fall as distinct ice needles. Sleet is partially melted snow. Drizzle is the term applied to small droplets ($< 500\,\mu$ radius) whilst rain consists of larger droplets.

Figure 2.18 indicates the process of hail formation in a *model* of a cumulo-nimbus cloud. The agglomeration of ice crystals, white in colour because of the entrapped air, falls through the supercooled water zone becoming coated with a layer of clear ice. The hailstone can then be swept aloft again in the hail

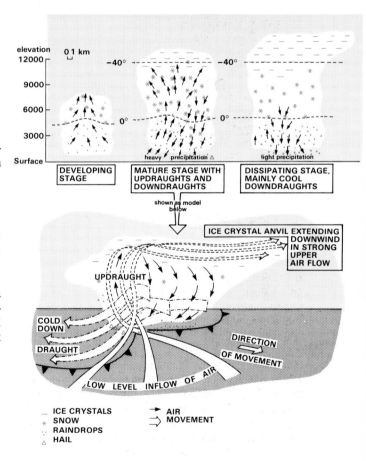

2.18 (above) A cycle of thunderstorm development and (below) a model of a mature cumulo-nimbus cloud

tower where velocities exceed 30 m/s and the process repeated. This leads to the build up of large hailstones of alternate clear and opaque ice layers which eventually fall from the cloud. The largest hailstone recorded weighed 680 g. Given the strength of our skulls it is fortunate that they are usually considerably smaller! Nonetheless hail damage to growing crops is a hazard in areas such as temperate continental interiors where pronounced summer convective heating occurs.

In addition to the formation of hail Fig. 2.18 also shows a small-scale *overturning* of the atmosphere. The energy surplus of the surface is being carried aloft in the uplift currents and the latent heat of condensation released, while the cold downdraught represents the compensating return flow. We mentioned earlier the transfer of energy in the atmosphere, the cumulonimbus cloud is thus an important mechanism of this energy transfer (the vertical heat flux).

In addition to classifying precipitation by its form such as drizzle, hail and so on, it is also possible to classify according to its origin. Such a *genetic classification* reflects the way in which the air has been cooled. The model described above and in Fig. 2.18 indicates a *convective* process of uplift and cooling, resulting typically from intense day-time heating of the land surface. A similar but larger-scale example, is the tropical hurricane where a complex of cumulus cells produce heavy and prolonged rain. A less intense pattern of convectional rainfall occurs when cool moist air begins to move across a warmer surface, leading to the warming of lower air layers and uplift in thermals. The resulting precipitation cells drift downwind as they grow and decay. You might like to consider the lapse rates which would need to prevail for this to occur. Such convectional rainfall in general terms thus tends to be scattered in time and space and results in showery conditions.

The second broad genetic category is *cyclonic* precipitation when air is rising along frontal surfaces. This is dealt with in more detail in the section which follows on atmospheric motion. The third category is *orographic* precipitation which is induced by cooling associated with uplift over a relief barrier. If you look at relief and precipitation maps of the British Isles, for example, this association between relief and rainfall is quite clear. It is probably best, however, to consider relief as magnifying cyclonic precipitation, so that orographic rainfall is a component of the total rainfall experienced at some stations.

ATMOSPHERIC MOTION

If you imagine a heated room next to a cold room, opening a door between them would result in a circulation of air until the temperature differences had been eliminated. Standing in the door you would find low down a flow from cold to hot room, higher up a flow from hot to cold. This is a simple model of thermal circulation. If we think of the differences between sea and land surfaces and the mosaic of land surfaces themselves it is possible to conceive of patches of differential heating in the atmosphere. These may lead to the development of local winds.

Land and sea breezes

Figure 2.19A shows wind, humidity and temperature conditions at Porton. Notice the shift in wind direc-

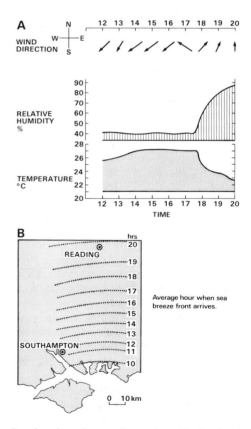

2.19 Land and sea breezes in southern England

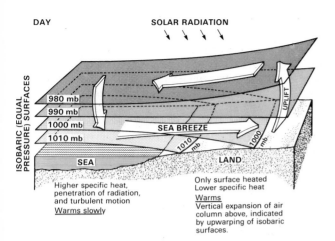

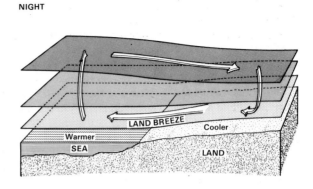

2.20   Land and sea breezes

tion at 1730 hours. What happens to the relative humidity and temperature as this shift occurs? Although Porton is 40 km inland these changes are typical of those which occur with the onset of the sea breeze. Figure 2.19B indicates the different times of day when the wind shift occurs as you move inland. Although existing on average only six times a year at Porton Fig. 2.19A does show that even in mid-latitude areas the *sea breeze effect* (temperature fall and relative humidity rise) can be detected. In the tropics the effect can extend up to 300 km inland with vertical depths of 2 km and be rather more frequent.

Figure 2.20 shows the processes at work. During the day solar energy cascades down onto the surface. On land only the surface itself is heated while at sea some of the sunlight penetrates so that a column of water absorbs the radiant energy. In addition to this there are considerable differences in specific heat, water needs to absorb five times as much energy to raise its temperature by the same amount as a comparable mass of land. Combining these facts means that the land warms up more quickly than the sea. The column of air above the land is heated and vertically expands. This is shown in Fig. 2.20 by the tilting of the *isobaric surfaces*, which represent lines of equal pressure in the atmosphere. The land surface develops a lower atmospheric pressure than the surface over the sea. Air will move from high to low pressure in response to this pressure gradient force. Therefore, as the diagram shows, a circulation of air develops from the sea at low levels and towards the sea at higher elevations.

At night the reverse situation prevails. The isobaric surfaces slope the other way, the land cooling more rapidly than the sea and a land breeze develops. This phenomenon is diurnal in nature and of course easily masked by more dominant movements in the atmosphere. Interestingly it is the only occasion when air moves directly from high to low pressure. As a medium, or *mesascale*, process land and sea breezes reduce the daily maxima and night-time minima of temperatures. The effects can be significant to man. In the case of Chicago on Lake Michigan the lake breeze reduces the lakeshore temperature maxima by as much as 16°C. The 7 m/s airflow in this example also makes the lakeshore more refreshing to live in than areas only a few kilometres inland. These continue to enjoy the undiluted pleasures of a Mid-Western summer, whose heat and humidity may be appreciated more by corn than by people!

Winds in the mountains

In many mountain areas anabatic and katabatic winds exist. The *anabatic* wind is an upslope flow developed during the day, the valley sides receive greater heating than the air at the same height above the centre of the valley. Air rises above the crests and feeds an upper return current along the line of the valley. At night a *katabatic* or downslope wind may occur when cold denser air at higher elevations drains down into the valley. Such a chilled downslope katabatic flow is particularly noticeable down valley from a glacier.

The forces acting on moving air

So far we have been looking at micro- and mesascales. When we increase the scale of our enquiry to involve distances exceeding ten kilometres and periods exceeding an hour we have to take into account the fact that any horizontal motion of air is taking place on a moving earth.

The primary cause of motion is the existence of a pressure difference, the *pressure gradient force*, introduced in the discussion of land and sea breezes. On surface pressure maps a close spacing of isobars indicates a greater force than a wider spacing.

The rotation of the earth's surface results in an apparent deflection to the right of the line of motion in the northern hemisphere and to the left in the southern. This *coriolis* force acts at right angles to the direction of air motion and is shown in Fig. 2.21. The motion of the air is parallel to the isobars, in other words at right angles to the pressure gradient.

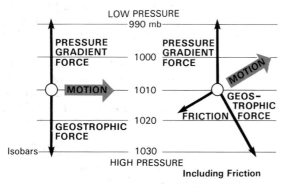

2.21   A vector diagram of balanced windflow in the northern hemisphere

Above about 1,000 m winds blow parallel to the isobars in this fashion, the *geostrophic wind*. At lower elevations surface friction lowers the wind velocity below its geostrophic value. Since the deflective coriolis force depends on velocity it too is reduced and at lower levels close to the surface winds actually blow across the isobars in the direction of the pressure gradient force as shown in Fig. 2.21.

The general circulation of the atmosphere

The very large-scale circulation pattern of the atmosphere is known as the general circulation. It is the mechanism by which the energy surpluses and deficits are balanced.

If we imagine initially a non-rotating earth there will be a pronounced thermal gradient in mid-latitudes, cold air lying polewards and warm air equatorwards. On a very much larger scale we would thus have a situation similar to the land and sea breeze process described earlier. Given the forces which act on moving air an attempt at equalisation will result in a simple westerly circulation in mid-latitudes.

This simple pattern is, however, distorted by the rotation of the earth and the existence of large mountain barriers such as the Rockies, which act as weirs in this flow. A series of large waves is therefore produced in this westerly flow. These usually number between four and six and extend through most of the troposphere. They are known as *Rossby waves*.

These waves vary in their seasonal position and amplitude. As they in turn guide smaller-scale and lower-level disturbances, they have a major effect on the weather and climate experienced at the surface. In Fig. 2.22 contrast the amplitude of the winter wave with the weaker summer wave. In winter cold arctic surface air moves south over the centre of the continent. In summer warmer and moister air moves across the centre of the continent. This is the *dynamic*

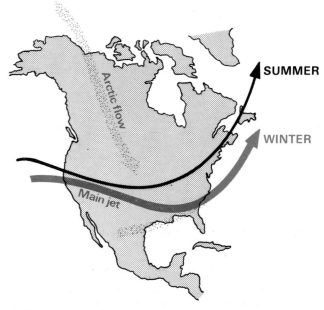

2.22   North America, upper air flows

explanation of the *continental interior climate* of the USA with its large yearly temperature range and summer precipitation maxima. These facts complement the differences between land and sea discussed in connection with land and sea breezes.

Within the Rossby waves, which typically have widths of 3,000 km and lengths of 6,000 km, we have a second-order wave pattern. Because such waves have sharper and tighter curves, centrifugal force outwards becomes important as it causes acceleration of the airflow on the ridge and deceleration on the trough limb. If you look at Fig. 2.23 Part I you will notice that in area A more air will be entering than leaving, resulting in *convergence* shown in Part II. In area B, where air is accelerating, more air leaves than arrives and we have *divergence*. This is occurring in the upper troposphere and the areas of convergence and divergence bring into being compensating circulations at the surface beneath the wave. Part III shows how such areas of high and low pressure are produced at the surface.

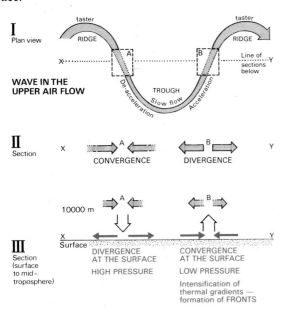

2.23 Convergence, divergence and upper air flows

The depressions of mid-latitudes

If these disturbances in the upper air westerly flow have a wavelength between 500 and 1,500 km they are unstable and the wave will amplify. As indicated in Fig. 2.23 convergence at the ground is produced at **B**. This convergence of air intensifies thermal gradients to such an extent that belts of rapid temperature change may be identified. These *fronts*, boundary surfaces in the lower troposphere, are thus produced by the general circulation. Since the disturbance aloft in the upper westerlies travels along the line of the Rossby waves so do the surface-level fronts and low-pressure areas or depressions.

Figure 2.24 shows a model of a depression with its series of superimposed airflows. At the 500 mb pressure level we have the cold dry airflow of the upper waves. At the surface in the west we have a broad airflow a few hundred kilometres wide and several kilometres deep flowing ahead of and parallel to the surface-level cold front. This air rises over the warm front, turning south as it climbs. This particular flow has been likened to a conveyor belt and represents *slant convection* carrying sensible heat (the warm air itself) and latent heat upwards. The circulation in the depression can thus be seen as an essential part of the general circulation. Slant convection is transferring energy from lower latitudes and elevations to higher latitudes and elevations.

The *Bjerknes Polar Front Model* of depression development was produced before knowledge of the upper air flow became available. It is therefore no longer satisfactory as an explanation of the origin of depressions although at the descriptive level it is still

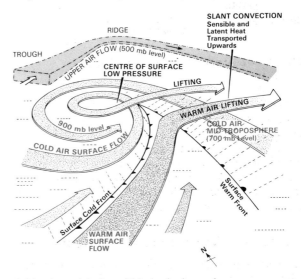

2.24 A model of a mid-latitude depression

useful in illuminating the growth and decay sequence. Figure 2.25A shows this sequence. It begins with the existence of the polar front, a lower tropospheric surface sloping north at a gradient of about 1:100, which separates tropical air moving from the west and polar air moving from the east. Corrugations in the front amplify as they move from west to east and at the bulge of the wave a low-pressure area develops and deepens. A circular pattern of isobars develop and the front becomes differentiated into warm and cold fronts. The term warm front simply means that an observer at the surface would experience warm air replacing cold as the travelling depression moved past.

The mature stage of the depression is shown in Fig. 2.25B together with its cloud sequences. A mature depression typically has a diameter of 1,700 km and a life cycle of between four and seven days. The final stages are characterised by *occlusion* when the more rapidly moving cold front catches up with the warm front and lifts the warm sector above the surface. The low-pressure area begins to fill and the depression dies. On the long trailing cold front shown in Fig. 2.25 another wave may have developed which could go through the same sequence. We may therefore find a 'family' of depressions strung out along the polar front.

The mid-latitude circulation is therefore characterised by an upper air westerly flow, within which are a series of long waves. These Rossby waves have within them a series of secondary waves which travel from west to east. The airflow in these in turn produces convergence and divergence and, at the surface itself, areas of high pressure (anticyclones) and low pressure. We have dealt with the latter in the section above and will return to them in the section on the weather and climate of the British Isles when the significance of the anticyclones will also be examined.

The general circulation in low latitudes

We can model the circulation in the tropics and subtropics by invoking the image of a thermally driven cell. At the equator we have a net radiation surplus and high temperatures near the surface. This produces a low-pressure area, the inter-tropical convergence zone (ITCZ), within which air slowly rises (*i.e.* 1–2 cm/sec) leading to condensation and precipitation. At about the 200 mb level the rising air diverges and begins to flow polewards. At the tropics the sinking limbs of the cell give subsiding air and stable conditions, beneath which air we find the subtropical anticyclones. At the surface a pressure gradient exists

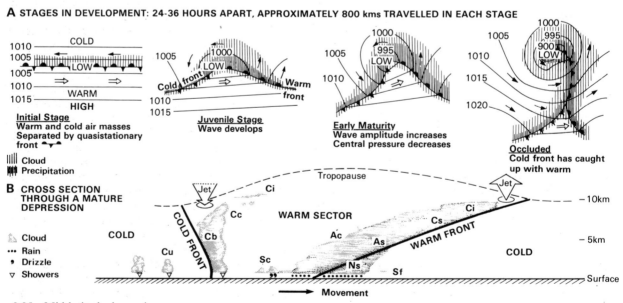

2.25 Mid-latitude depressions

between these high-pressure cells and the ITCZ resulting in the formation of a lower-level return flow equatorwards.

The image created so far is one of a large thermally driven cell, termed the *Hadley cell*. In fact, of course, with the forces which act on moving air (see Fig. 2.21) the meridional, or low to high latitude, transfer is subordinated to air motion parallel to latitudes. Thus the upper air flow has a westerly component, the lower tropospheric flow an easterly component. At the surface itself because of friction the easterlies flow with a component directed towards the low pressure, forming the *NE Trades and SE Trades* in the northern and southern hemispheres respectively.

The low-latitude circulation therefore involves upcurrents in the ITCZ. This converts the latent heat acquired by the surface trades in their passage over the warm tropical oceans into sensible heat and cumulus clouds.

The general circulation: global winds and ocean currents

Early models of the general circulation invoked images of three cells, the primary cell being the subtropical Hadley cell. A second thermally driven cell was believed to exist in high latitudes where cold subsiding air moved equatorwards with a return flow polewards aloft. The third cell was envisaged as being driven indirectly by the other two in mid-latitudes. This is no longer acceptable and Fig. 2.26 attempts to show the surface winds and upper air flows. This is a complex diagram and will repay careful study. The low-latitude Hadley cell may be recognised (see the sections at the edge of the maps), although the symmetry of one either side of the equator may exist only in the autumn and spring. The second major point is the existence of horizontal mixing in the mid-latitudes westerly surface wind belt and the return flow of the Hadley cells. The latter takes place in troughs extending from the mid-latitude westerly belt into low latitudes on the western sides of the subtropical high-pressure areas.

Figure 2.27 indicates how the energy is transferred. The sensible heat transfer is accomplished by wind, and some 20% of the total is transported by the oceans. The latent heat transfer curve is rather different: in low latitudes this heat transfer is towards the equator! This occurs because of the convergence of the trades

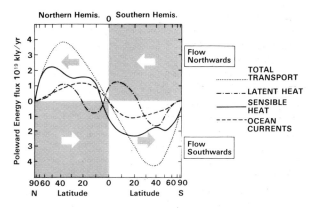

2.27 Meridional heat fluxes

towards the ITCZ. These winds pick up energy and this latent heat is released as sensible heat with condensation in the rising air of the ITCZ.

The *global wind belts* shown schematically in Fig. 2.26 can be interpreted in the context of the general circulation. The trades, which actually blow over 50% of the globe's surface, originate on the margins of the subtropical high-pressure cells and converge on the ITCZ over the oceans. In the summer hemisphere and over continental surfaces the zone between the two trade wind belts is characterised by the equatorial westerlies. Over the Atlantic and Pacific Oceans the ITCZ doesn't shift poleward enough for this westerly flow to develop and the convergence zone has been termed the Doldrums, a region of light and variable winds so feared in the days of sailing ships.

The mid-latitude westerlies, sometimes known as the Ferrel westerlies, emanate from the poleward side of the subtropical high-pressure cells. They are in fact quite variable as a result of the disturbances, *i.e.* depressions, which travel through them. They are most pronounced in the southern hemisphere where land and sea differences are less disruptive of the general flow. Finally there are the polar easterlies. Since the idea of a third cell in high latitudes is now suspect these are best viewed as easterly winds on the poleward side of the mid-latitude depressions.

Oceanic circulation is controlled by the low-level winds (*i.e.* by the subtropical high-pressure cells and the westerlies) together with the configuration of the ocean basins. Equatorwards of the cells the trades generate the north and south equatorial currents. On

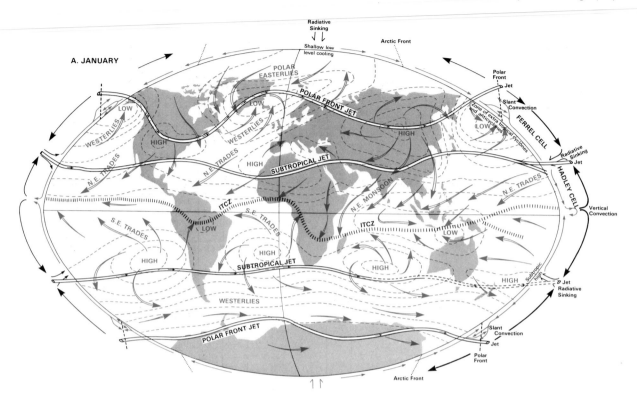

2.26   The circulation of the atmosphere: January and July

the western side of the oceans these swing polewards around the western margins of the cells. In the case of the Atlantic, for example, this water can be imagined 'piling up' in the Gulf of Mexico, where it becomes warm and stable, before escaping between Florida and Cuba as the Gulf Stream. Equivalent currents of high-velocity, poleward-directed warm water are the Kuro Shio, Brazil, Mozambique, Agulhas and East Australian currents.

On the poleward side of the high-pressure cells westerly currents dominate, the West Wind Drift in the southern hemisphere and the North Atlantic Drift and North Pacific Drift in the northern. On the eastern side of the oceans, currents swing round the easterly sides

of the high-pressure cells, for example, as the Canary and California currents in Atlantic and Pacific respectively. Such currents are broad flows of low velocity and with the deflection away from the land upwelling of colder water from below occurs.

Ocean currents can thus be viewed as a consequence of the general circulation and the energy budget of the earth. The magnitude of their effect in transporting energy and the general 'lag' effect of the oceans on global climate is not yet clear in detail. The effect of the warm North Atlantic Drift on climate in the British Isles will be mentioned below while Fig. 2.28 indicates the interactions in southern Peru. Subsiding air of the subtropical high-pressure cell gives intense

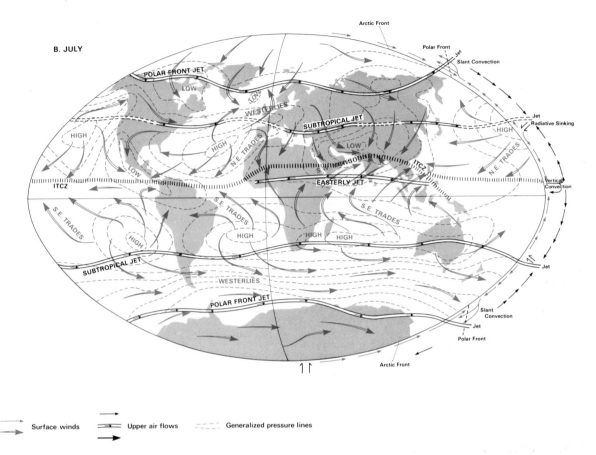

B. JULY

Surface winds ━━▶ Upper air flows ═══▶ Generalized pressure lines ─ ─ ─

aridity to the Atacama Desert. The easterly tropical airflow results in an upwelling close to the coast with colder deeper water replacing the 27°C surface water. The air immediately above the cool sea is chilled producing a temperature inversion. At shallow elevations, *i.e.* less than 800 m, a sea breeze develops as the land is warmer than the sea. This breeze, the Garua, drifts inland carrying the mist produced by the sea's cooling effect. Once inland the mist disperses as the air is warmed by the land. Such situations of coastal aridity are also found in south-west Africa and Morocco. In southern California the inversion is an element in the air pollution problems of the Los Angeles Basin which were depicted in the Introduction.

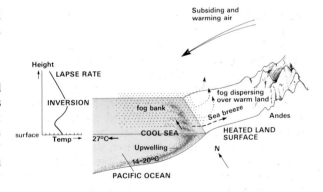

2.28   The west coast of South America

REGIONAL EXAMPLES OF WEATHER AND CLIMATE

The British Isles

The weather and climate of the British Isles results from a complex interaction of such factors as the general circulation, the travelling depressions, the distribution of land and sea and ocean temperatures.

Although they vary their position and intensity, the Icelandic Low and Azores High are permanent features of the mean surface pressure map. The Icelandic Low as a mean feature merely reflects the high frequency of depressions in the area. In summer the circulation is weaker, depression frequency is less and their general tracks lie further to the north. In winter an additional factor is the occasional incursions of the Siberian High into western Europe. This is a shallow but intensely cold high-pressure area which when it extends westwards can divert depressions from their normal tracks, either northwards via the Norwegian Sea or southwards to the Mediterranean.

In winter temperatures at the surface are often more than 11°C warmer than the latitudinal average. This positive temperature anomaly reflects the existence of the *North Atlantic Drift*, although its effects on temperatures and precipitation must take into account the motion of air across it. The water in the Drift actually takes nine months to arrive off Ireland and another three months to reach Norway. This 'liquid Florida sunshine' will release its energy to arctic and polar air flowing southwards which will gain sensible and latent heat as they cross the Drift. If air flows northwards, from the Azores High, it will of course be cooled in its lower layers.

A useful concept when describing the dynamic climatology of an area is the *air mass*. Imagine areas where air rests quietly. Over a period of time it will acquire the temperature and humidity characteristics of the surface beneath. Such source regions typically occur in anticyclones where large-scale subsidence exists. The source region, a sea surface of relatively uniform temperature or a land surface of uniform temperature and dampness, imparts its characteristics to the overlying air. The air mass is thus a large body of air in the lower troposphere which has uniform temperature and humidity in a horizontal direction. The properties of the air mass do not only depend on the nature of the source region. As the air mass moves it is modified by the nature of the surfaces which it passes over, a process influenced by the length of time its journey takes.

Air masses may be classified into Polar or Tropical and subdivided according to whether they have a continental or maritime source, *i.e.* mP, cP, mT and cT. In the case of the British Isles the most common air mass is mP, with a source region close to Greenland. This can only approach across the ocean and, according to the season and the track it has followed, has different modifications and characteristics.

A surface pressure situation with a ridge of high pressure extending northwards from the Azores towards Iceland and a low over the Baltic will result in a northerly mP airstream affecting the British Isles. With lapse rates close to the DALR to some height, this air will be outstandingly unstable. Over the sea heavy cumulus will form and showers will affect land exposed to the north. Inland conditions will be clear and cold at night, by day showery and with temperatures cooler than average.

mP air can also approach on a north-westerly track, for example streaming south-east behind the cold front of a travelling depression. Such air may have had a long history originally starting as cP over Canada and acquiring maritime characteristics over the Atlantic. Streaming from the north-west towards the British Isles such air is often known as mPK, K indicating its cold nature. mPK is moist in its lower layers which will also have been heated as it crosses the North Atlantic Drift. This will result in unstable conditions: abundant cumulus and showers which intensify as the airstream passes over higher ground in the north and west. Inland it may well bring bright cool showery days and clearish cold nights.

In the warm sectors of depressions we may find mP air which has approached after a long arcing passage over the Atlantic. This is known as mPW air (*i.e.* W = warm). It is a variable air mass. Travelling northwards, its lower layers are being cooled and thus stability is increasing. It carries considerable moisture, in winter tending to give stratiform cloud (St or Sc) as a result of its stability. Forced uplift over hills may result in heavy precipitation which will of course augment precipitation associated with uplift along frontal surfaces. In summer mPW may well give clear weather but with its high moisture content strong land heating will cause instability and the birth of cumulus and showers.

mT air reaches the British Isles in the warm sectors

of depressions forming well to the south. At its source, in the subtropical high-pressure cell, it has high temperatures and humidity. Tracking northwards its lower layers become cooled and characteristically give stratiform cloud cover, drizzle and mild temperatures in winter. In summer, of course, heating of the land surface breaks up this cloud cover inland.

During winter and spring outbreaks of cP air may affect Britain. With its source in northern Russia this air is dry, cold and stable. In its short track across the North Sea the lower layers acquire heat and moisture which may bring snow showers to eastern districts. Inland it may give characteristic clear, cold and sparkling winter anticyclonic weather. Even more infrequent is cT air from the Sahara approaching from the south-east. If it reaches Britain its lower layers are stable and it gives hazy, cloudless and dry weather.

The above discussion of air masses and weather types is highly generalised. The Monthly Weather Record details daily weather patterns. Examining these for particular summer and winter months (*e.g.* Fig. 2.38) will complement the picture sketched above,

especially when used with the information of mean situations available from atlas maps of pressure, winds and precipitation.

A case study of the 1975–6 drought

The sixteen-month period of precipitation deficiency which ended in September 1976 was unprecedented since weather recording began in Britain in 1727. Figure 2.29 indicates the distribution of this precipitation deficiency. Abnormally high surface pressure existed near southern Britain during 1975–6. What effect did this have on the deficiency pattern over the country? Much of the discussion in the section above evoked the movements of depressions and their associated surface air streams. As we will see in this section the other element in the climatology of the British Isles is the existence of the high-pressure anticyclones.

Like many phenomena the drought did not have a single cause and as with many natural systems it was influenced by a number of feedback factors. During 1971–4 winters were mild in Europe and the major

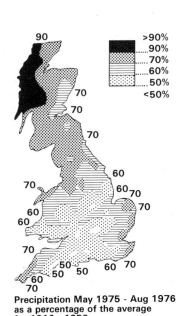

Precipitation May 1975 - Aug 1976
as a percentage of the average
for 1916 - 1950

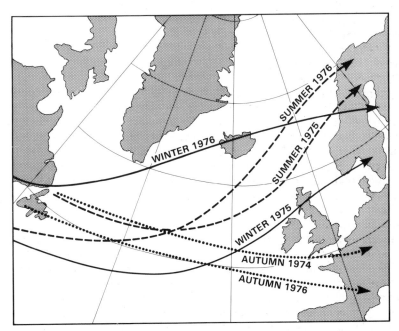

POSITION OF JET STREAM, 1974 – 1976

2.29   The drought of 1975–6

focus of northern hemisphere cold was displaced eastwards to the northern Pacific and Canada. During the actual period of the drought the jet stream over the eastern Atlantic was displaced 5–10° further north compared to its mean position in the period 1951–70. Figure 2.29 indicates the position of the jet which guides the movement of surface-level depressions. This displacement could have been caused by the warmer winters of 1971–4 which, as suggested above, resulted in the Pacific Ocean north of 40° being cooler than usual, especially by the spring of 1975 and through 1976. This coldness produced a stronger temperature gradient over the area and an enhanced upper air westerly flow. Bearing in mind the large-scale wave patterns mentioned earlier such enhanced Pacific flow results in a deeper wave over Canada and more intense winter cold. This disturbance in the wave pattern will have an effect 'downstream' resulting in an anomalous ridge extending from the south over Europe.

This circulation pattern with a *blocking anticyclone* extending as a surface ridge from the Azores High resulted in far fewer cyclonic disturbances, and hence frontal rain, passing over the British Isles. The position of the upper air westerly jet in Fig. 2.29 reflects this situation which was unusually sluggish in reverting to the more normal patterns.

The exceptional rains which followed the drought owed their origin in part to this unusual summer. As a result of the hot summer experienced in the ridge of high pressure, sea temperatures were 2°C warmer than usual off southern Britain. This allowed more moisture, in absolute terms, to be present in the overlying air. When the blocking anticyclone finally decayed to be replaced by an upper air trough the south-westerly surface air flow was able to bring this ample moisture to the 'thirsty' earth of southern Britain.

Weather and climate in the Mediterranean Basin

As can be seen from the temperature and precipitation data shown in Fig. 2.30 the climate of the Mediterranean is characterised by hot dry summers and mild wet winters. A closer examination, however, shows considerable variations reflecting the 3,000 km distance from west to east and the interdigitation of sea and land.

As a result of its position between 30° and 45°N the area's weather is controlled by a westerly airflow in winter and by the subtropical anticyclone in summer. The existence of mountain barriers inhibiting a north–south exchange of air are also of some significance.

In winter, with the sun overhead in the southern hemisphere, the mid-latitude westerlies are displaced southwards. However it is too simple to assume that depressions simply enter from the west to produce the Mediterranean Basin's winter rainfall. Only about 9% of the winter depressions originate in this way. Far more significant are the *lee depressions* which form behind the Atlas, Pyrenees and Alps; as can be seen from Fig. 2.30. The lee depression occurs when a westerly airflow is forced over the relief barriers. As the air is forced over the mountains there is a lateral expansion and divergence. On descending there is lateral contraction and convergence of air. This results in the formation of wave troughs which may develop into closed low-pressure systems. The lee depression is therefore different in origin from depressions which form as a wave along a frontal zone (shown in Fig. 2.25).

Figure 2.30 also shows in the eastern Mediterranean the existence of a frontal zone, where temperature differences of up to 16°C are found between Mediterranean air and cT air from the Sahara. This temperature contrast gives energy to cyclogenesis, so that the winter rainfall of this area is associated with depression development and movement. Finally, the fact that sea temperatures are warm in winter is of some importance when discussing precipitation origins. Convective instability exists, especially when mP air enters the basin.

Although the winter season is associated with depression development in fact they do not dominate the circulation. Depression development is rare under *high index* situations. The term high index is applied when the upper air westerly flow is displaced polewards, without any large-scale waves in it. The term *low index* is applied when there are large waves in the upper westerlies, *i.e.* a blocking anticyclone at 20°W may result downstream in a pronounced north–south flow across western Europe. In this low-index situation depression birth in the lee of the relief barriers in the western Mediterranean will be favoured. The surface weather during the winter in the Mediterranean may therefore vary with the *index cycle*. Although it is the moist season there may be relatively few rain days.

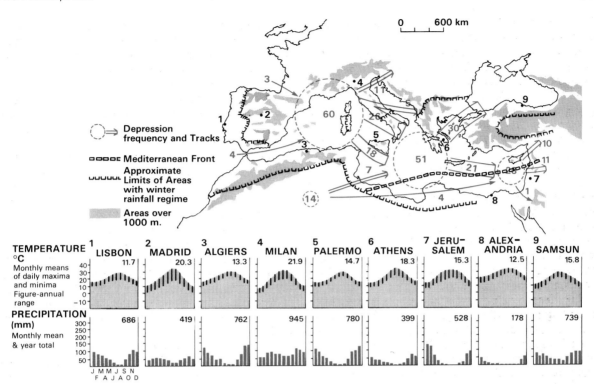

2.30   The climate of the Mediterranean Basin

During the spring conditions become somewhat unpredictable. The Eurasian winter high-pressure area decays and a discontinuous north-eastwards extension of the Azores anticyclone begins, associated with a northwards shift in the westerly airflow and its associated disturbances.

The expanded Azores anticyclone dominates the basin during the summer. The period of summer dryness varies from about two months in the north to between eight and nine on the African coast. The air at low levels is moist and solar heating is strong, but aloft there is the usual pronounced subsidence in the subtropic anticyclone resulting in stability. An additional factor explaining the low precipitation in summer is the absence of any cold air. Thus fronts and cyclogenesis do not occur.

A number of *local winds* reflect the existence of the relief barriers and their associated cyclogenesis. The Mistral is a cold northerly wind flowing through the lower Rhône valley. It is best developed when a depression lies in the Gulf of Genoa and a high-pressure ridge from the Azores anticyclone lies over the eastern Atlantic. Under these conditions mP air is drawn down the relief funnel of the Rhône valley where there are associated katabatic effects. The Bora is a similar fierce cold wind in the northern Adriatic occurring when air streams through the gap between the Alps and the Dinarics behind an easterly moving depression. These winds are of considerable local significance to agriculture. Winds from the south also move northwards ahead of easterly moving depressions, *i.e.* cT air from the Sahara. These are known as the Scirocco in Algeria, the Leveche in Spain and the Khamsin in Egypt.

The Mediterranean basin is thus characterised by variability: seasonal variability resulting from the apparent migration of the sun, winter season variability associated with index cycles and depression birth and movement. The appreciable size of the basin and the interpenetration of land and sea together with appreciable relief barriers also contribute to the spatial variations.

Hurricanes

Tropical cyclones, known as hurricanes in the Atlantic and typhoons in the Pacific, are the single most impressive atmospheric phenomena. Occurring in late summer and autumn they originate over the western sections of the tropical oceans between 5° and 15° north and south of the equator. Figure 2.31 is a model of a mature hurricane. If you look at the airflow you will see that the hurricane is basically turning the atmosphere upside down.

The hurricane begins as a shallow wave in the tropical easterly flow, which may develop into a tropical depression and storm. A hurricane only develops from these disturbances when high pressure exists aloft. This is vital, as the divergence associated with such a high-level anticyclone allows the surface low-pressure area to deepen and be maintained.

The structure of the hurricane is indicated in Fig. 2.31. At the centre is the eye, typically 20–50 km across, within which there is subsidence, with adiabatically warming and cloud-free air. The eye is surrounded by an amphitheatre of towering cloud systems reaching to the tropopause. Wind speeds in the vortex around the low-pressure area are high, up to 360 km/hr, and this increases the transmission of sensible and latent heat from the warm ocean surface to the atmosphere. Within this ring of cloud, *uplift* produces condensation and heavy rain, as much as 500 mm/24 hours. The release of latent heat with this condensation produces the warm core to the vortex, which intensifies the upper anticyclone so that the *outflow* of air aloft is maintained. This outflow permits the low-level *inflow* of heat and moisture to continue, as long as the moving hurricane remains over a warm ocean, whose surface provides energy in the form of sensible and latent heat. Friction over the sea is also low and the winds spiral around, rather than towards the centre of the low, which thus retains its intensity.

The low-level influx of heat and moisture is also reflected in the structure of the hurricane away from its centre. Bands of cloud spiral into the low, as can be seen in Fig. 2.31 and Fig. 2.32. These spirals are often about 80 km apart at the edge of the disturbance, but thicken and close up as the centre is approached.

A hurricane at sea would involve the following sequence. The day before the storm's arrival would be fair and calm. A long sea swell, formed by storm waves outrunning the storm, would be accompanied by a cloud sequence similar to that marking the approach of a warm front in a mid-latitude depression. Rain showers would develop, the pressure fall

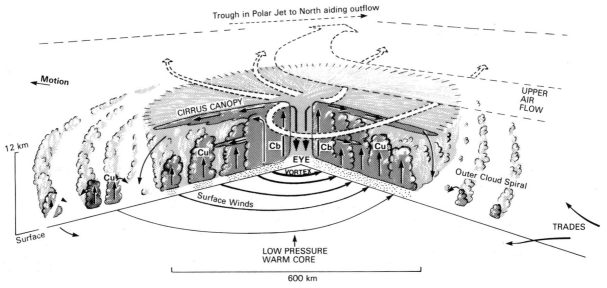

2.31   A model of air circulation in a hurricane

and wind speeds increase. The full fury of the storm would arrive with high winds, rain and heavy cloud.

Depending on the speed of the hurricane several hours of this intense storm would be followed by a clear sky and almost calm air as the eye passed. The sea, of course, would continue to be very rough and the cloud, rain and winds would commence again as the other wall of the eye approached and passed. The wind direction would be reversed and gradually, with rising barometric pressure, weather and sea conditions would improve.

Fortunately there are on average only 60 hurricanes a year. Their tracks are shown in Fig. 2.34. They usually last two to three weeks over the sea, moving with the upper easterly air flow before shearing polewards. Once over land, the absence of the warm ocean's sensible and latent heat reduces the energy supply and the hurricane decays. Increased friction over the land also produces a centripetal air flow which begins to fill the low.

The hurricane is an *energy transfer mechanism*, with its heated surface air rising to the troposphere and cold upper air descending in the eye. It is thus a significant part of the general circulation. They are also unpleasant natural hazards for mankind. The wind-driven waves batter coastlines, the high winds damage structures and the heavy rainfall produces floods. Hurricane Diane in August 1955 killed 200 people and damaged $1,500,000 worth of property as it swept through the eastern seaboard of the USA and mention has already been made of the massive loss of life in Bengal.

2.32   Hurricane Gladys, 8th October 1968

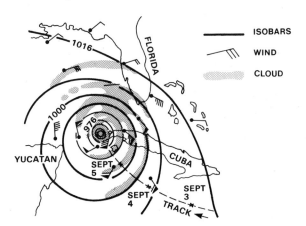

2.33   Surface weather map of a hurricane

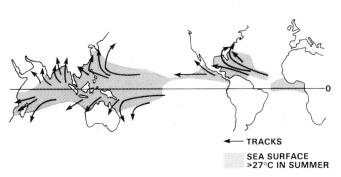

2.34   Principal hurricane tracks

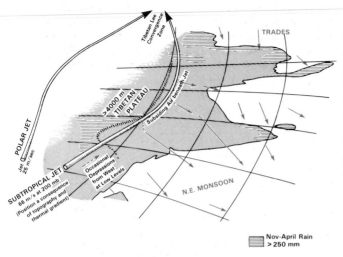

2.35   The Indian Monsoon: winter airflow and precipitation

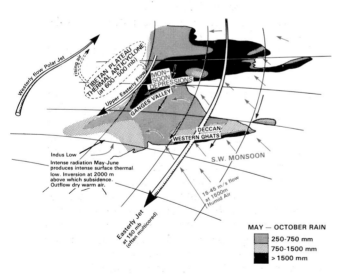

2.36   The Indian Monsoon: summer airflow and precipitation

The Asian monsoon

The graceful dhows of the Arabs made use of large-scale reversals of surface winds in the Indian Ocean for their trading. Their word *mausin* (season) has given its name to the monsoon. In this region we have the familiar pattern of the subtropical highs, the trades and easterly waves, and also a unique arrangement of mountains which inhibit north-south air movements. The Asian monsoon is in fact the result of an *interaction of planetary and regional factors* at the surface and aloft. The simplistic view of the summer SW monsoon over India as a large-scale sea breeze, and the winter NE monsoon as a land breeze linked with the movement of the overhead sun, ignores these factors. It is not intended in the brief section below to describe the climate in detail, merely to portray the large-scale circulation patterns.

Figure 2.35 shows the winter air circulation. At the surface the winter outblowing NE monsoon is associated with fine sunny weather over most of India. Aloft the westerly jet is split, flowing either side of the 4,000 m high Tibetan plateau. The southerly branch is positioned as a result of a pronounced temperature gradient and the existence of the relief barrier to the north. As Fig. 2.35 attempts to show, air subsides beneath this jet, giving surface high pressure and outblowing winds in the winter dry season.

Between March and May the upper westerlies begin their seasonal migration polewards and the northerly branch of the jet strengthens. Conditions at the surface are still dry but with warming temperatures. Usually in late May the southerly jet begins to break down erratically and eventually all upper air westerly flow is diverted north of the Tibetan massif. Figure 2.36 shows the situation in summer. The Equatorial surface trough of low pressure has moved north, as the sun is now overhead at the Tropic of Cancer. To the south of this, at about 15°N, we now have a high-level easterly jet. As this high-level easterly flow develops, at the surface humid south-westerly air encroaches more and more over the subcontinent. Its arrival, the *burst* of the monsoon, marks the end of the hot dry season and the start of the period of heavy monsoon rain. In spite of the fact that India receives 80% of its rainfall in this period, there is no simple weather pattern. If you examine pressure, wind and rainfall maps from an atlas these variations will be apparent. For example, the dry area in the north-west where monsoon air is overlain by subsiding continental air leading to an inversion, little convection and thus little rain occurs. In peninsular India the higher rainfall on the western Ghats is apparent. The south-west monsoon flow from Madagascar is shallow, 1,500 m in thickness, picking up moisture on its long ocean track. The convective instability of this air is only released by forced uplift over the Ghats with a noticeable rain shadow in their lee. Thirdly, if we look at the Ganges valley, monsoon depressions occur moving towards the west with a frequency of about two a month. Without these the rainfall here would undoubtedly reflect relief conditions far more.

In the autumn with the southward retreat of the Equatorial trough the summer circulation pattern decays. By October the easterly trades affect the Bay of Bengal at the 500 mb level generating disturbances at their confluence with the retreating equatorial westerlies. These disturbances can develop into the hurricanes (cyclones) which affect the area. By October the westerly jet is re-established south of the Himalayas and the winter circulation pattern exists again.

The large-scale circulation pattern outlined above is subject to some variability. There are variations in the date of arrival of the south-westerly surface airflow; the failure of the monsoon to arrive 'on time' has severe human consequences. Within the summer period itself breaks can occur when the mid-latitude westerlies occasionally push south of the Tibetan barrier.

CONCLUSION

In this chapter various processes and patterns in the atmosphere have been introduced. The treatment of necessity has been selective and more detailed descriptions can be found in the further readings.

Although the treatment has been selective inspection of Fig. 2.37 will show that you have been introduced to a range of processes operating at different temporal and spatial scales. This atmospheric motion (kinetic energy) is a response to the way in which energy flows through the atmosphere. Early in the chapter the global energy budget was identified with its surpluses in low latitudes and at low elevations and its deficits in high latitudes and altitudes. The convective cumulo-nimbus cloud, the hurricane and the mid-latitude depression can be seen as mechanisms

working to correct this imbalance. At a range of scales therefore weather and climate can be viewed as a response to the existence of the sun as the sole energy source for atmospheric processes and patterns. Finally it should be mentioned that the topic of urban climates is examined in Chapter 7.

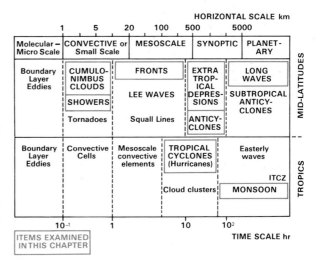

2.37 Scales of atmospheric motion

Review Questions and Exercises
1. Describe and account for the pattern of solar radiation shown in Fig. 2.8.
2. Explain what happens to solar radiation as it passes through the atmosphere and reaches the earth's surface.
3. Figure 2.9 showed solar radiation receipts for slopes of different aspects. Taking a topographic map (1 : 25,000) of an area of dissected terrain, use a grid (*i.e.* 250 m) and assign an aspect and slope inclination for a sample of squares. Using Fig. 2.9 determine the maximum amount of solar radiation, the hour in which it occurs and estimate the total daily radiation for the winter and summer solstices. Describe the variations you have discovered and speculate on the importance of variations in slope aspect and angle in influencing ground temperatures and cultivation.
4. How and why do clouds differ in form?
5. Under what conditions and for what reasons do the various types of precipitation occur?

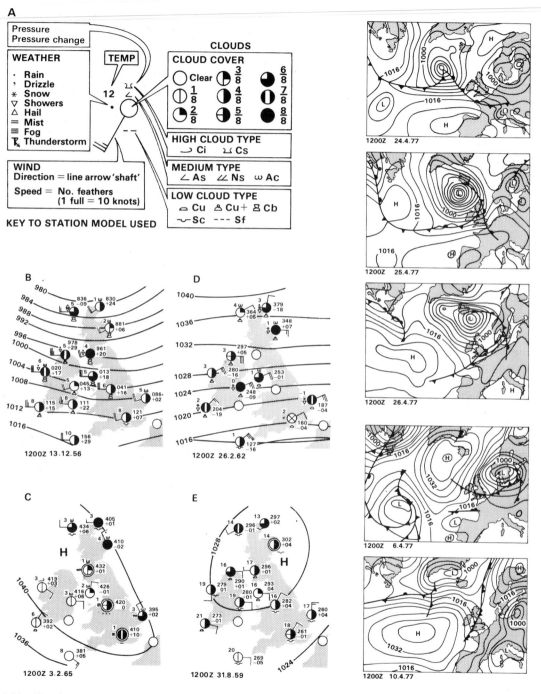

2.38 Weather maps

2.39 North Atlantic pressure

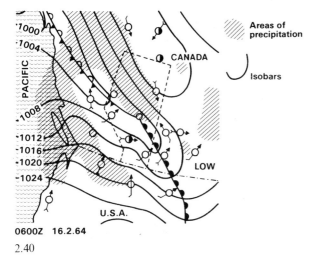

Areas of precipitation

Isobars

PACIFIC

·1000
·1004
CANADA
·1008
·1012
·1016
·1020
LOW
·1024
U.S.A.

0600Z 16.2.64

2.40

6. What are the causes and effects of atmospheric stability and instability?
7. In what kind of situation, and for what reasons, would you expect land and sea breezes to develop?
8. Describe the internal workings and associated weather patterns of *a*) mid-latitude depressions and *b*) hurricanes.
9. Describe the pattern of planetary wind systems and ocean currents. What part do they play in maintaining the energy budget of the earth?
10. Using the key in Fig. 2.38A describe and account for the weather conditions experienced in the various parts of the country under the conditions shown in Fig. 2.38B, C, D and E.
11. Figure 2.39 shows an April Atlantic surface pressure situation and the location of fronts.
    a) Comment on the 24–26th April situation. What weather sequences would you expect in Penzance, Rekjavik and Stockholm?
    b) Contrast the wind, cloud and temperatures experienced over Britain on 6th and 10th April.
12. Figure 2.40 is a February weather map for Western Canada. The low pressure is drawing Pacific air across the Rockies and a chinook (föhn) wind descending in the lee of the Rockies is raising temperatures up to 25°C. Why?

Further Reading

Barry, R. J. & Chorley, R. J., 1976, *Atmosphere, Weather and Climate*, Methuen.

Boucher, K., 1975, *Global Climate*, English Universities Press.

Barrett, E., 1974, *Climatology from Satellites*, Methuen.

Calder, N., 1974, *The Weather Machine*, B.B.C.

Chandler, T. J. & Gregory, S. (Eds.), 1976, *The Climate of The British Isles*, Longman.

Flohn, H., 1969, *Climate and Weather*, Weidenfeld & Nicolson.

Hanwell, J. & Newson, M., 1973, *Techniques in Physical Geography*, Macmillan.

Harvey, J., 1976, *Atmosphere and Ocean: Our Fluid Environments*, Artemis Open University.

Lockwood, J., 1976, *World Climatology: An Environmental Approach*, Arnold.

3 Weathering and Slopes

INTRODUCTION

In a cemetery in West Wilmington, Connecticut, P. Rahn made a simple examination of the extent of weathering on tombstones, whose date of erection could be determined from their inscriptions. His students divided the weathering of headstones into six classes:

1. unweathered
2. slightly weathered (slight rounding of corners and letters)
3. moderately weathered (rough surface)
4. badly weathered (letters difficult to read)
5. bad weathering (almost indistinguishable letters)
6. extreme weathering (no letters left)

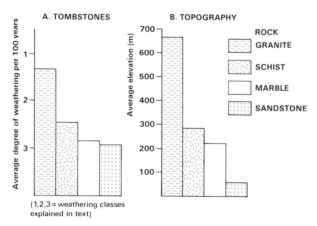

(1,2,3 = weathering classes explained in text)

3.1   Rock type, tombstone weathering and topography

Figure 3.1A represents the average degree of weathering of the four types of rock used for these tombstones, the sandstones having weathered most rapidly and the granite least. Figure 3.1B also shows that average elevation in this area varies with rock type. We might conclude then that differences in rock hardness, and their resistance to the effects of the atmosphere are therefore significant factors in determining a landscape.

The earth, of course, has been 'out of doors' a lot longer than these Connecticut tombstones. In this chapter we will be looking at the processes of rock decay and the formation of slopes. The chapter therefore begins to link the initial chapters on the lithosphere and atmosphere and build towards our understanding of the processes of landscape development.

Since the entire landscape is made up of slope elements it is not surprising that slope origin, form and development forms a crucial area of earth science. However, as Kirk Bryan wrote in 1940, the geomorphologist after long discussion 'wakes with a gasp to realise that in considering the important question of slope he has always substituted words for knowledge, phrases for critical observation'. Since these words were written there have been advances in our understanding of slope processes but slope problems in general still present some of the most fundamental questions in geomorphology.

Geomorphologists have studied rock weathering and slope development with theoretical formulations, with field investigation and with laboratory experiments. In this chapter we will look at all three approaches and see how they help us interpret landforms and understand the supply of material to erosional systems.

ROCK CLASSIFICATION

In broad terms what happens at the earth's surface depends on three variables: climatic conditions; a range of local variables such as exposure, vegetation and the presence of water in the ground; and the properties of the material involved. This latter involves the composition of the rock and the existence of voids and planes of weakness which determine its susceptibility to weathering. It is therefore appropriate to start our treatment with an examination of rock character.

We will be using the term rock in its broadest sense meaning any substance of the mantle or crust. Minerals are any naturally occurring inorganic substance with a definite chemical composition. Since there are several thousand varieties of minerals a detailed treatment is beyond the scope of this book, but the minerals present in a rock determine its properties.

Rocks are conventionally divided into three broad groups according to their origin. The first of these are the *igneous* rocks which have solidified from molten magma. *Extrusive* igneous rocks are those which have been ejected onto the earth's surface where they have cooled rapidly either as lava flows or as ejecta—the dust, ash and cinder produced when there is considerable gas present in the magma. Of the lava flows basalt is perhaps the commonest, occurring as vast lava plateau spreads and associated with sea-floor spreading. On cooling the dense black basalt shrinks giving characteristic columnar jointing.

In contrast to the extrusive igneous rocks such as basalt, andesite and rhyolite, the *intrusive* or plutonic rocks have solidified more slowly at depth. Their texture is thus different from the glassy or fine-grained texture of the extrusives. Mineral crystals up to 150 mm across can exist which interlock to give a dense strong rock. The composition of the commonest intrusives is given in Fig. 3.2.

Sedimentary rocks form the second major group. *Clastic sediments* consist of particles derived from a previously existing rock which have been weathered, moved and finally deposited. After settling the particles become hardened, by pressure or compaction driving out water, or alternatively by cementing. Cementation is achieved by slowly moving ground waters which import the cementing material as ions in solution; typically this cementing mineral may be silica (SiO_2) or calcium carbonate $CaCO_3$). This accumulates in the pore spaces between the particles

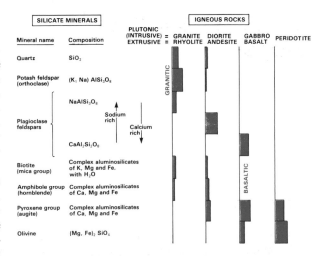

3.2   The composition of igneous rocks

of original rock but rarely fills them completely, the remaining voids being filled by water or by hydrocarbons such as oil and gas. Such sediments deposited and compacted or cemented in layers (strata) often have bedding planes separating successive layers. Therefore in considering their breakdown we have to consider the nature of the original particles, the nature of the cement and the existence of voids and bedding planes.

The processes of transport tend to sort the particles according to their size and density and thus clastic sediments are conventionally divided according to the nature of their constituent grain sizes. Conglomerate represents gravel bars or beaches which have become lithified (*i.e.* changed to solid rock). The pebbles (larger than 2 mm diameter) are rounded in a conglomerate; in contrast a breccia consists of angular fragments. Sandstones, where the grains typically lie between 0·06 and 2 mm diameter, consist of grains of any durable mineral, the commonest being quartz. Where the finest particles, the clays and muds, have been compacted, shale results. Shales are mechanically weak and are easily split into flakes, this fissile quality renders them geomorphologically weak.

In addition to the clastic sediments there are also sediments produced by the growth activities of animals and plants. *Limestones* consist of the limey parts of such organisms as corals, algae, foraminifera, clams, snails, etc., which have accumulated in layers to be later compacted. As silica, in the form of quartz grains or chert, and clay minerals can be present, limestones cover an enormous variety of chemical and physical properties. Its colour can range from white (*e.g.* chalk) to black, its texture from granular to dense, and its general appearance from the massively jointed carboniferous limestones, to those which are barely altered fossil reefs. As the term limestone merely indicates that more than 50% of the rock is calcium carbonate ($CaCO_3$) it is therefore important to determine the real nature of the rock when considering its weathering and landform producing qualities.

Many limestones have the mineral dolomite (calcium magnesium carbonate) incorporated with the calcite, where its proportion exceeds 50% the rock is known as *dolomite*. Since this mineral is not produced by organisms as shell material it is possible that the process of dolomitisation reflects magnesium ions from sea water being substituted for part of the calcium ions. Finally, where the sea or lake water at the time of deposition was heavily charged with lime in solution, direct chemical precipitation could form the soft lime mud known as *marl*.

Geologically of considerably less importance than the limestones but of considerable economic significance are the *hydrocarbon* compounds in sedimentary rocks. Vast bog accumulations of partly decayed plant matter (peat) may under compaction become lignite, coal and anthracite as progressively more water is removed.

The final group of sedimentary rocks are the *evaporites* which represent salts precipitated from shallow desert lakes or constricted lagoons. They are mainly sulphates of calcium (gypsum, anhydrite) or sodium chloride (halide or rock salt) and are often interlayered with shales and limestones representing fluctuating environmental conditions at the time of their formation.

The third major group of rocks are the *metamorphics*. These are rocks which have had a change in their mineral state in response to a change in their environment, typically high temperatures and pressures associated with orogeny and intrusion (see Chapter 1). The original minerals have been recrystallised, new minerals have been formed and new structures imposed, obliterating, for example, the original bedding structure in sedimentaries. Metamorphic rocks, produced therefore from both igneous and sedimentary rocks, are generally harder and more compact than their original parents.

Considering the fine-grained, clay-rich, marine clastic sediment known as shale, with squeezing and shearing it is transformed into slate, a denser fine-grained rock which splits into sheets along cleavage lines distinct from the original bedding surfaces. With greater pressure and shearing, phyllite is formed which breaks into curvy sheets and, under extreme pressure, schist which has a foliated structure.

Conglomerates, sandstones and siltstones can be metamorphosed to produce quartzite. Under pressure the quartz grains are crushed and forced into closer contact and a silica cement from the migration of underground waters fills the pores between grains. Unlike its parents, quartzite doesn't fracture around the grains, it will break across them. It is a hard and resistant rock.

During the processes of internal shearing involved in metamorphism the calcite mineral in limestone is re-formed into larger and more uniform crystals, bedding planes are obscured and mineral impurities are drawn out as swirls of colour. Marble is thus the metamorphosed equivalent of limestone.

Gneiss is the general term applied to a metamorphic rock showing banding, the elongation of crystals. Granite gneiss, for example, results from the flowage of granite in a plastic state. It is a massive rock, retaining a similar texture to granite but it has a streaky appearance representing the elongation of crystals. The metamorphism of igneous rocks and the metasediments mentioned above is complex and beyond the scope of this treatment.

Figure 3.3 indicates the cycle of rock transformations and the linkages between the three broad categories of igneous, sedimentary and metamorphic which have been introduced here.

WEATHERING

Weathering might be defined as the decomposition or disintegration of rocks *in situ* at or near the earth's surface. *In situ* doesn't imply static, there will be losses of some minerals, other may be gained and some moved about, but all this will occur at the same site. It is often convenient to consider two broad groups of weathering processes, physical (or mechanical) weathering where there is no alteration of the existing minerals, and chemical weathering where some of the rock-forming minerals are altered or decay. Since in nature these rarely occur in isolation it is perhaps more

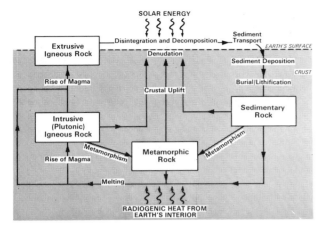

3.3   The cycle of rock transformation

helpful to think in 'terms of processes leading to disintegration and decomposition rather than rigidly apply the chemical and physical division mentioned above.

Disintegration

Disintegration of rock results from insolation, biological agents such as burrowing animals and plant roots, and the growth of salt and frost crystals. In general disintegration occurs without a significant alteration of minerals and such agencies are known as mechanical or physical weathering agents.

Frost action is perhaps the simplest of these to understand, burst water pipes in underinsulated British homes being a graphic demonstration of the fact that ice occupies 9% more volume than an equivalent mass of water. In rocks which contain many small fissures the disruption by phase change (*i.e.* from water to ice) can be considerable. Laboratory experiments indicate that in a closed system (*i.e.* where there is no communication with the outside) a pressure of up to $2,100 \, kg/cm^2$ is produced at $-22°C$ by the freezing water. A closed system may be rare in nature, not all the crevices and pore spaces may be filled with water, but water in a crack will commence freezing from the surface and conditions deep within the crack may approach those of the closed system. Since the tensile strength of rocks is well below the theoretical pressure generated by the water–ice transformation (*i.e.* granites $70 \, kg/cm^2$, limestones $35 \, kg/cm^2$ and

sandstones $7–14 \, kg/cm^2$) the existence of frost wedging is not in doubt. The litter of granular debris beneath cliffs in arctic and alpine areas and the existence of *felsenmeer* (a sea of angular rocks on horizontal surfaces) are testimony to the efficiency of this process (Fig. 3.4).

When porous rock saturated with water falls below freezing point ice crystals begin to form in the larger pore spaces. The crystals develop in clusters at right angles to the freezing surface and they attract water from the smaller pores around them. This selective enlargement of ice crystals means that the coarse pore space is filled with ice. These crystals continue to grow, drawing supercooled water from adjacent pores, and break the rock when the pressure of their growth exceeds the strength of the porous material.

Frost action therefore consists of two processes, wedging and inter-pore crystal growth. Earlier it was mentioned that the nature and effectiveness of any process depended on three variables: climate, the properties of the material involved and local factors. In the case of frost action susceptible material would be rock with fissures or with ample pore space. In the case of climate the requirements are the existence of water and the frequency of temperature fluctuation around $0°C$. Such climatic conditions are found in mid and high latitudes and altitudes, in autumn and spring when large temperature ranges exist, often not just diurnal but also associated with frequent alternations of cyclonic and anticyclonic weather. Finally, the local variables include the degree of exposure and the

3.4   Frost riven boulders (felsenmeer) Glyder Fach, Wales

absence of vegetation, which would influence the depth to which temperature changes could penetrate.

A process similar to that of ice-crystal growth can occur with the evaporation of water from the surface of a rock whose pore spaces are filled with a saline solution. The evaporation of the solvent, water, leads to the growth of *salt crystals* which can disintegrate the rock. Such a process can only occur on weaker rocks which are already partly weathered and where the moving ground waters can dissolve the salt which is carried to the surface as the water evaporates. A semi-arid climate with considerable evaporation is obviously called for. Preventing the upward migration of saline ground waters into buildings, with the possibility of disintegration of the structure, is a design consideration in such areas. In the natural environment the existence of niches in exposed rock walls (Fig. 3.5) is an example of the process and its resulting form.

Biological agents such as vegetation roots and burrowing animals may also physically disintegrate rock. In the case of the former, however, it is difficult to separate the physical wedging of a growing root from associated chemical attack on the rock.

The existence of spheroidal boulders (Fig. 3.6) and of granular disintegration of boulders composed of a variety of minerals has often been cited as evidence for *insolation-induced weathering*. In the case of the spheroidal boulders, rocks were envisaged as being poor conductors of heat. Exposed to the sun the outer layers expanded, pulling away from the cooler core. At night the process was reversed, a process known as exfoliation. In the case of granular disintegration different minerals with different co-efficients of expansion responded to day-time solar heating and nocturnal cooling by breaking the rock into its constituent minerals. It all sounds so logical, but there is clear evidence that even in conditions of extreme aridity and intense solar heating it is chemical processes dependent on the presence of water which are predominant in producing this kind of rock breakdown. Observations on fallen granite columns in Egypt show that it is the shaded and buried faces which are most weathered, not those exposed to the sun. Laboratory experiments subjecting rocks to repeated heating and cooling cycles also fail to replicate rock breakdown until moisture is present.

On a much larger scale to spheroidal boulders a similar pattern of fractures concentric with the rock surface can be seen, resulting in exfoliation domes up to several thousand metres in extent. This phenomenon is particularly noticeable on granite rocks and has been attributed to pressure release. The rock was originally formed at depth. With the removal of overlying material, the rock expands and dilates, producing *sheeting* of the outer rock layers.

Decomposition

Rock decomposition or chemical weathering consists of a range of processes involving hydration, hydrolysis, oxidation, carbonation and solution. In broad terms these result in increases in bulk of the minerals,

3.5   Natural arches, an extreme development of weathering niches, Utah

3.6   Spheroidal weathering in olivine basalt, Antrim. (Not all of this form of weathering is produced in deserts!)

decreased density and decreased particle size. Igneous and metamorphic rocks having been formed under conditions of high temperature and pressure are subject to chemical changes progressing towards the creation of minerals which are more stable under conditions existing at the surface. The general order of mineral stability is:

	Ferromagnesian Series	*Plagioclase Series*
LEAST STABLE	Olivine	Calcic plagioclase
	Augite	(e.g. anorthite)
↓	Hornblende	Sodic plagioclase
	Biotite mica	(e.g. albite)
MOST STABLE	Orthoclase feldspar	
	Muscovite mica	
	Quartz	

With this order of susceptibility it is not surprising to find quartz and muscovite as common constituents of rocks which are composed of weathering residues.

The role of water in rock breakdown is fundamental. In hydration, for example, water is absorbed into the crystal lattice of the minerals but there is no fundamental chemical change. It can, however, cause swelling (in some shales as much as 60%!) and the disintegration of rocks following alternate wetting and drying. One of the unpleasant effects of the 1976 drought in Britain was in fact the shrinkage of clays leading to partial foundation failure in many parts of the country.

In the case of *hydrolysis* there is a chemical union of the water with the minerals of the rock. The H^+ ions displace metal cations in the silicates and the OH^- ions combine with these to form solutions which are carried away. High concentrations of H^+ ions cause silica (SiO_2) and alumina (Al_2O_3) to become linked to form complex clay minerals. In the case of granodiorite rock, for example, the mineral orthoclase is acted on by carbonic acid and water:

$$2KAlSi_3O_8 + H_2CO_3 + nH_2O \rightarrow K_2CO_3$$
$$+ Al_2(OH)_2Si_14O_{10} . nH_2O \div 2SiO_2$$

producing soluble potassium carbonate, the clay mineral kaolinite and soluble silica. Such reactions are complex and rather beyond our scope here. It is important to note, however, that the amount of water (*i.e.* the availability of H^+ ions from the rain and soil water) affects both the type and the degree of weathering.

Oxidation occurs when oxygen ions combine with metallic ions, the oxygen being readily available in water. This process is most apparent in rocks which contain iron in sulphide, carbonate and silicate forms. In the case of olivine the process starts with hydrolysis:

$$MgFeSiO_4 + 2HOH \rightarrow Mg(OH)_2 + H_2SiO_3 + FeO$$

The ferrous iron then reacts to form limonite:

$$4FeO + 3H_2O + O_2 \rightarrow 2Fe_2O_3 . 3H_2O$$

In connection with sedimentary rocks oxidation is important in weathering clays like montmorillonite where O combines with the Mg and Fe ions.

The reactions outlined above are simplified and of course cover only a limited range of examples. Later in the chapter we will examine the weathering of limestone and tropical weathering in more detail. At this stage it is important to note that the processes of hydration, hydrolysis, oxidation, carbonation ·and simple solution usually occur in conjunction. The effects of reactions with water, oxygen, iron and carbon dioxide is basically to weaken the structure of rocks and change their volumes. Outside of the humid tropics typical end products of chemical weathering are sand and clay. In the humid tropics both these are unstable and compounds rich in aluminium, iron and water form the waste (the laterites to be mentioned later).

INVESTIGATING WEATHERING PROCESSES AND FORMS

Screes or talus slopes

Where freely exposed steep rock slopes occur, particles loosened by weathering cannot remain still, they fall to accumulate at the base of the cliff as talus. Normally they pile up to form a slope typically between 25° and 40° depending on the type of rock, the character of the constituent particles, the friction binding the particles and the operation of supply and removal processes. Although they occur in all climates wherever free faces (cliffs) are found they are especially evident in glaciated highlands where overdeepened valleys provide many free faces and frost weathering produces an abundant supply of debris (Fig. 3.7).

Statham (1973) describes an investigation of scree slopes which illustrates one way in which *theoretical, field* and *laboratory* work can be combined to build towards an understanding of an earth surface process and its resulting forms.

On the basis of a review of earlier work he wrote 'screes commonly have characteristics which are not fully explained by a model of debris avalanching at a constant angle of repose'. The characteristics he referred to were the *sorting* of debris on the slope (with larger material at the foot) and the fact that scree *profiles* consist of an upper straight slope, often below the maximum angle of stability for the material, and a lower slope with some concavity.

He began his work with a *theoretical* discussion. If at first the free face is high, particles landing on the initial debris slope have high energy and they could bounce, roll and slide to form a long tailing concavity. As the scree piles up, the headwall (or free face) becomes shorter and material falling from it has less energy than in the initial case. Also, as the length of the scree slope increases, only high energy particles could travel as far as the concave section. This particular slope element could therefore be imagined as getting smaller as the height of the free face becomes less. The concave element would be progressively replaced by a straight slope, where particle supply is balanced by the loss through impact and debris avalanching.

Statham secondly considered the theoretical movement of particles. Assuming a supply of mixed sized particles falling onto a debris slope with uniform particle size distribution over its total length, he proposed a sorting mechanism. The smaller falling particles couldn't travel very far down the debris slope because of high frictional losses. They could thus accumulate at the top, whilst larger fragments could move further down the slope. Statham also reasoned that this process was self-propagating—once smaller particles predominated at the top of the scree, the 'trapping

3.7   Scree slopes beneath frost riven cliffs, Alberta, showing the profile and size sorting discussed in the text

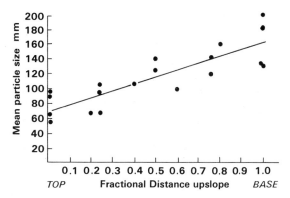

3.8   Cader Idris screes: particle size and distance upslope

depressions' in its surface would be smaller, and only material smaller than this could accumulate. Meanwhile the larger debris could continue to roll further down the slope. In this way size sorting of rock debris on the scree could occur.

He then attempted to confirm this with a simple *laboratory experiment*. This consisted of two boards covered with angular debris of different sizes (11–13.5 mm and 22–27 mm). Placing a pebble on the boards he tipped them until the pebble moved and noted the angle. This was repeated fifty times for a range of pebble sizes. This simple experiment confirmed the theoretical assertion that large pebbles are capable of moving greater distances downslope than smaller particles. Retardation started when pebbles met material slightly smaller than itself and particles tended to stop in material of its own size.

It was only at this stage—after the literature review and the theoretical treatment—that Statham subjected his ideas to *field testing* to discover if the scree profiles and sorting postulated and demonstrated in the laboratory existed in the field. He investigated a number of screes in the Cader Idris area, measuring their slopes and sampling the size (and shape) of material. According to his theoretical formulations the importance of the basal concavity decreased as the height of the headwall decreased. In the field, however, it proved impossible to accurately measure the height of the headwall! This illustrates a fundamental field problem in geomorphology—not everything we would like to measure can in fact be measured.

Figure 3.8 shows particle size against proportional distance upslope. The regression is significant at the

0.0001 level. The field evidence thus confirmed the theoretical and laboratory work which had suggested an increase in particle size downslope resulting from variations in the rate of deceleration. Rockfall-controlled screes (*i.e.* those with no removal of material from the base) therefore are not merely simple angle-of-repose slopes.

In our second example we will again look at what happens on a scree slope. Although the scree slope appears simple, like the rest of the environment, it is in fact quite complex. One way to attempt to understand it is to isolate it and simplify it in a way which emphasises its structure and the web of relationships which exist. Figure 3.9 represents this *systems analysis* approach to a scree slope.

Initially we might look at the input and output of material on the scree, the solid line. Debris falls from the cliff on to the upper scree slope where it either rolls and slides to the lower scree or is trapped on the upper slope and stored before creeping to the lower scree slope. In turn, on the lower scree slope where some vegetation exists, the debris may be trapped or creep further downwards to leave the scree subsystem and enter the stream channel system. This path, shown on Fig. 3.9 with conventional systems notation, is a

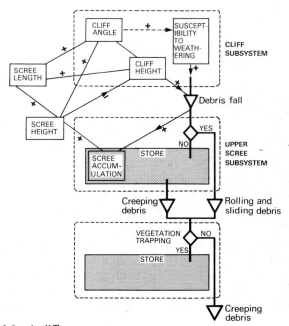

3.9   A cliff-scree system

cascade. *Cascading systems* are thus defined either by the path of mass, or energy. You will also notice that the scree has been simplified and divided into three subsystems which are interrelated. Finally, of course, the output from the scree system (the debris from frost shattering) is in turn the input to the sediment transport system of the streams.

Figure 3.9 also shows a *morphological system*, the network of relationships which exist between different forms in the system. The direction of connectivity is shown by the arrows—for example, as scree height increases cliff height decreases. The strength and direction of connectivity is usually revealed by regression analysis.

The morphological system depicted in Fig. 3.9 also shows the existence of a *feedback loop*. Look at the debris fall. As this increases, scree accumulates, leading to an increase in scree height which in turn leads to a reduction of cliff height and a reduction in debris fall. This is a negative feedback, which leads to self regulation or self repairing—a kind of dynamic equilibrium. Such negative-feedback loops are very common in almost all systems in the natural environment. As we will see later in the book, if man fails to understand the operation of natural systems he can seriously affect these self-repairing, conservative feedback loops.

Geomorphologists are of course interested in the interlocking and mutual adjustment of cascades and morphological structures. Figure 3.9 taken as a whole therefore represents a *process-response system* and it shows the manner in which form relates to process. Although the scree slope example discussed above is a relatively simple example it illustrates the simplifying and explanatory function of systems analysis. It also introduced you to cascading, morphological and process-response systems, which will be useful to you in later sections of the book.

Chemical weathering: the solution of limestone

As a result of the dominance of the solutional process limestone areas are noted for their lack of surface water and a range of distinctive landforms such as dry valleys, surface depressions and cave systems. Such distinctiveness soon attracted geomorphological explanation. Much of the early work on limestone landforms, however, was often qualitative in nature

and expressed in a 'cyclic' framework, involving the description of a supposedly evolutionary sequence in the development of the landscape. More recently limestone areas have attracted attention because measuring the rate of the solutional process is relatively simple. The section which follows illustrates the results obtained by such detailed and quantitative studies of process.

Limestone may be defined as any rock which consists of more than 50% calcium carbonate ($CaCO_3$), a broad definition within which considerable variations of texture, jointing, strength and composition exist. Much of the discussion which follows is concerned with Carboniferous Limestone, a massive well-jointed and strongly bedded rock. It is important to remember however that solution is equally important in weaker rocks such as limestone shales and chalk.

Pure water at 10°C (a typical ground water temperature in Britain) is capable only of a modest amount of solution of $CaCO_3$, approximately 12 parts per million of calcium (ppm). Water with an addition of carbon dioxide (CO_2) forms carbonic acid which acts on the $CaCO_3$ to produce calcium bicarbonate which is carried off in solution:

$$CaCO_3 + H_2O + CO_2 \rightleftharpoons Ca(HCO_3)_2$$

Where is this CO_2 coming from? Simplistic explanations point to the atmosphere. As we know from Chapter 2 the atmosphere contains a small amount of CO_2, typically 0.033% by volume. If this is dissolved

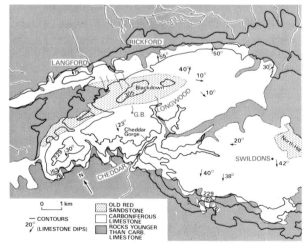

3.10 The Mendip Hills

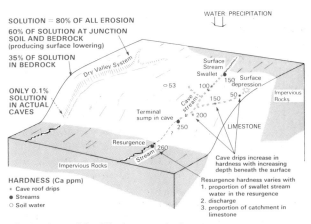

3.11 A model of limestone solution

clines which dip away at either end of the fold axis rather like the keel of an upturned boat. The centre of these periclines is composed of Old Red Sandstone, flanked by younger rocks, the lower limestone shales and the Carboniferous Limestone series.

This area has been the most intensively studied limestone area in the world; more than 10,000 water samples have been taken over a fifteen-year period. Figure 3.11 attempts to summarise this work. It is a model, or simplification, of the geological and topographic setting of central Mendip and the location and magnitude of the solution process.

It is interesting to imagine our understanding of limestone solution in system terms. There is an input—water and CO_2—and an output—Ca in solution. What occurs in between was initially unknown to us and can be called a 'black box'. What has happened is that, as a result of many people's work in the laboratory, on the surface and within the caves, we begin to have a more accurate idea of where and how the Ca is removed. This partial understanding might be called a 'grey box'. If we await the evolution of particularly thin and sinuous cavers and the development of robust yet accurate instruments, the hazy detail of the grey box may become clearer to us and be transformed into a

in rainwater it is capable of dissolving limestone to produce a concentration of 74 ppm at 10°C. Yet water hardness in limestone areas is often four times this figure! Obviously additional CO_2 must be available, either within the soil (*i.e.* the weathered mantle) or rock.

Measuring the concentration of CO_2 within the soil atmosphere is far from easy, in Britain during summer under permanent pasture it can reach 1·6%. If all this gas was dissolved in the soil water it would be ultimately capable of producing a Ca concentration of 280 ppm, a figure close to that measured in many limestone springs (resurgences). This CO_2 is called biogenic CO_2 and is obviously important in limestone weathering.

Figure 3.10 shows the topography and geology of the Mendip Hills in Somerset. Topographically central Mendip is a plateau (230–260 m) flanked by steep slopes to north and south. The plateau is underlain by a series of east–west trending periclines, anti-

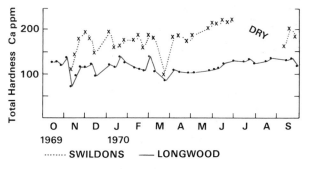

3.12 Water hardness in swallet streams

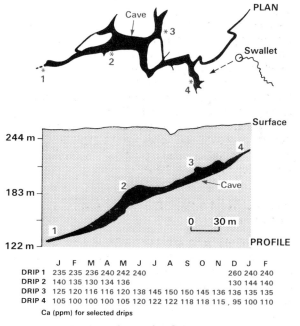

3.13 Water hardness figures for G.B. cave

'white box' when all the subsystems are known.

Figure 3.11 contains a range of detail about the location of limestone weathering in this model environment. Rainwater seeping across the surface of the impervious Old Red Sandstone forms surface streams which may disappear into a swallet once they have passed on to the limestone. In the model this water is indicated as containing 150 ppm of Ca. Figure 3.12 is an example of the results of detailed water sampling in this situation. The location of the streams is shown in Fig. 3.10.

Rain falling on the limestone itself passes into the rock, making use of the myriad fissures in the joints and bedding planes. This may eventually join the stream water as drips from the cave roof. Figure 3.11 attempts to show that this water increases in hardness with greater thickness of overlying limestone. Figure 3.13 is an example of detailed observations in GB cave on which the generalisation in the model is based.

3.15   Limestone pavement above Malham Cove, Yorks.

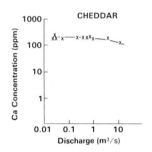

3.14   Resurgence discharge and water hardness

By the time the water emerges at the resurgence it is effectively saturated, carrying as much Ca as it can. There are, of course, variations at this point in the system; one example is shown in Fig. 3.14, which is a solution rating curve for Cheddar. You will notice from this that as the discharge increases (reflecting more rainfall, for example) the concentration of Ca expressed in ppm decreases. If you think about this it doesn't mean that the actual amount of Ca removed per day drops, just its concentration in a larger amount of water. It is often useful, therefore, as you will see later in the book, to look carefully at such figures and think about what they really mean.

In summary it is important to note the left-hand side of the model. Contrary to what might seem obvious at first sight only 0.1% of solution occurs in the cave systems themselves. The majority of solution in Car-

boniferous Limestone areas is taking place at the interface between the soil and the bedrock.

Having outlined the process of limestone solution it is now appropriate to look at some of the landforms produced. Cave systems, although spectacular, are relatively insignificant in the total picture. Figure 3.11 indicated that most solution occurred at the soil/rock junction and the removal of Ca produces over a period of time a lowering of the surface. By measuring the water hardness and the amount of water flowing from a known area, estimates can be made of such surface lowering. In the Mendip area estimates of

3.16   A surface depression developed on Carboniferous Limestone, Mendip Hills, Somerset

38–81 $m^3/km^2/pa$ have been produced, equivalent to a rate of surface lowering of 38–81 mm/1,000 years.

In particular locations where the overlying soil has been removed, limestone pavements can be seen (Fig. 3.15). The surface of the limestone has been etched, pitted and transected by grooves and flutings ranging from several millimetres to several metres in depth. The pavement is developed across the bedding plane and the vertically arranged flutings represent jointing weaknesses. The crevices are often filled with residual insoluble debris. Although exposed to our view in only a few locations the etching of the rock surface is evidence of the effect of the weak carbonic acid produced within the soil. These pavements are known as *lapiés* or *karrens*.

Surface depressions, ranging in diameter from tens to hundreds of metres, are another distinctive feature of limestone areas. Quite varied in form, they range from the swallet (or pot hole), associated with the disappearance of a surface stream, to the far more numerous closed depressions (Fig. 3.16). There are two views on their origin—solution from the surface downwards or depressions resulting from the collapse of a cave passage beneath.

In both cases we need to ask ourselves what the detail of the process might be. Limestone is both porous and permeable. Porosity measures the void space, ranging between 1% for massive limestones to 50% for corals, and permeability (the rate of transmission of water through the rock). Secondary porosity, or the interconnected void space, is a more significant term to apply to water movement through limestones. When this movement is intergranular and through small fissures it is very slow, and most solution will be concentrated near the soil/rock junction. Gradually the crevices widen from the small 10–20 μ width to about 5 mm, a process which may take 50 million years. Once a crevice exceeds 5 mm or so a *threshold* is passed, flow of water ceases to be slow and the solutional rate increases sevenfold. The first crack to reach this threshold thus starts to enlarge at a much faster rate than its neighbours. This process occurring at the surface leads to an irregular lowering of the ground level, *i.e.* with more solution at certain points, and the development of closed depressions. Alternatively the enlargement of some fissures into caves may occur at depth. Ultimately the overlying material fails, producing a collapse feature.

Closed depressions indicate a fundamental problem in geomorphology … *equifinality of form*. There may be two processes: solution from the surface downwards with concentration on master fissures, or collapse over a void. Once formed, the movement of residual material down the sides of the depression may result in their shapes being essentially similar. In other words two distinct origins may produce the same form. We will return to this general problem later in the book.

Before leaving weathering a dimension to introduce is the problem of variations resulting from *climatic differences*. Normally the rate of a chemical reaction increases with increasing temperature. Between the latitudes 30°N and S continuous high temperatures, frequent rainfall and the existence of relatively luxuriant vegetation has encouraged widespread deep weathering of silicates, the commonest rock-forming minerals. These break down into a regolith, or weathered mantle, consisting of sand (quartz) and clay (dominated by kaolinite). This is highly permeable and allows the chemical alteration of fresh rock to persist to depths of 200 m or more. The details of this tropical weathering will be examined later in connection with tropical soils and vegetation.

It has been argued that limestones present a situation where this normal picture of increased chemical weathering in tropical environments no longer holds true. The basis for this assertion lies in the fact that cold water is capable of absorbing more CO_2 than warmer water. This would mean a greater potential for attack on the limestone by the carbonic acid so formed and of course a greater concentration of Ca being carried off in solution. However, paradoxically tropical environments show many well-developed solutional landforms.

There are a number of explanations for this mismatch. Temperature is only one of the factors controlling the rate of limestone solution, the availability of water is another. In arctic and alpine environments water may exist in the liquid unfrozen state for only part of the year. In the mid-latitudes and tropics it may exist throughout the year and the total solutional loss over time may be greater. The role of biogenic CO_2 has been stressed earlier; in cold environments vegetation is sparse compared to the rates of growth and decay found in tropical environments. These two factors—water availability and biogenic CO_2—may thus offset the effect of temperature when the world scale pattern of limestone solution is examined.

Finally there is the problem of climatic change. In

3.17   Stonebarrow Down. A vertical aerial photograph, 10th August 1969

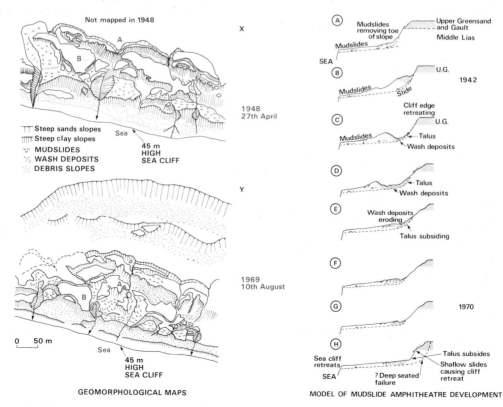

GEOMORPHOLOGICAL MAPS

MODEL OF MUDSLIDE AMPHITHEATRE DEVELOPMENT

3.18   Mass movement on a Dorset cliff

the tropics limestone areas have been subjected to un-disturbed weathering over long periods of time. During the Pleistocene, middle and high latitudes had their landform evolution disturbed by the direct effects of ice or by periglacial conditions. This latter term is applied to environments beyond the limits of direct glaciation where the existence of frost action and melt-water has influenced the landscape (see Chapter 6).

In the light of the complications hinted at here it is obvious that the degree of weathering revealed in the relative freshness or maturity of landforms is a very imperfect guide to the current rate of operation of the particular process.

MASS MOVEMENT

A cliff slope in Dorset

Our concern earlier in the chapter has been with weathering. It is now time to consider the movement of rock and regolith, the weathered debris at the surface. Figure 3.17 is a photograph of the coast at Stonebarrow Down, four kilometres west of Lyme Regis in Dorset. If you look carefully the lighter unvegetated areas are signs of two sorts of movement, massive rotational slides and a complex of mudflows.

This area has been chosen to introduce some of the processes of *mass movement*. This is a term which includes all downslope movements of rock and soil produced in response to gravity. It excludes movements when the material is carried by ice, snow, water and wind. The detail which follows may appear a little complex and the 'why' of the various processes will be examined separately later. It is important to realise, however, that mass movement often occurs as a combination of processes on landscapes which have three dimensions (not just a section up the slope!), and that this occurs over time.

This area has been studied since 1946 to discover the spatial and temporal patterns and magnitude of some mass-movement processes. Brunsden (1974) outlines the details of these studies which are particularly interesting as they extend over such a long period. Figure 3.17 shows the general form of the area. At the top of the photograph is the crest of Stonebarrow Down at an elevation of 150 m. Seawards from this is a cliff (up to 37 m high) developed on the Upper Greensand, beneath this is a sloping bench (the under-cliff) developed on the Middle Lias. This 200 m wide undercliff is bordered at the seaward edge by a 45 m high cliff. The Upper Greensand is a fine sand capped with thick chert beds and it weathers to individual grains of sand which 'run' when wet. The Middle Lias consists of closely fissured clays and silts with irregular limestone bands. It weathers to a fine silt clay.

Processes are relatively dynamic here and changes occur quite quickly. Under the attack of the waves the sea cliff itself retreats at an annual rate of between 1 and 5 m. Between 1942 and 1971 the crest at the top of the Upper Greensand retreated up to 60 m.

On 14th May 1942 a 'failure' occurred. The upper part of the 'cliff', together with a radar station located close to the edge, moved rapidly downslope. Figures 3.18A and B show the situation immediately before and after this event. Material was being continually removed from the undercliff by mudslides. The base of the upper part of the slope became more and more unsupported, failure occurred and a *rotational slide* of the Upper Greensand took place. Figure 3.18B shows the tilting of the Upper Greensand beds reflecting this rotational movement.

Figures 3.18C, D and E show the rotational slide element moving downslope, being reduced in size as weathering and erosion takes place. The freshly exposed Upper Greensand cliff (exposed when the

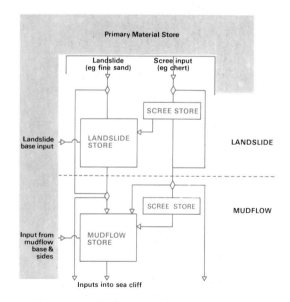

3.19 The Dorset cliff represented as a cascading system

failure took place) also weathers and the upper cliff edge retreats. The weathered material falls to accumulate as talus (similar to the screes already mentioned earlier in the chapter but consisting of smaller particles). Clay and sand are also washed down this slope by water and accumulate behind the barrier formed by the Upper Greensand landslide element (shown in the diagram as wash deposits). Figures 3.18F and G show this process continuing with the mudslides removing material across the undercliff. Figure 3.18H shows that conditions are again ripe for another deepseated failure.

Figure 3.18X is a map of the undercliff in 1948, 3.18Y in 1969. The large-scale elements, the rotational landslides, have moved down and across the undercliff and have broken into smaller units. Looking at such morphological maps, maps portraying the assemblage of landforms, emphasises the importance of thinking in three dimensions. Material on the undercliff is also moved by mudslides, these produce amphitheatre (*i.e.* arcuate) shaped embayments, some of which are drained by streams. Material is therefore removed from the sides as well as the base (*i.e.* over the edge of the seacliff).

Figure 3.19 shows this situation as a *cascading system*. Material from the primary store (the bedrock) enters as landslides or screes and also by erosion from the base of the moving landslides and mudslides. It is stored, for varying amounts of time, and leaves the system to form the input to the seacliff system.

Brunsden also gives details of some detailed surveys of processes at work on the undercliff for the period 1966–9 (Fig. 3.20A and B). Although many slides occur between January and March Fig. 3.20 shows how irregular the processes are and emphasises how important it is to monitor slope processes over a period of *time*. Figure 3.20C shows the number of slides by volume. The mean size was 4·1 m³. Is this figure that helpful, when as the graph shows, slides varied in volume from 0·02 to 86·1 m³ with most slides involving less than 1 m³ of material? Brunsden also calculated that the mean size of slide (the 4·1 m³) recurred every 1·6 months. Earlier in this discussion the major landslide elements from the upper part of the slope were examined and you might remember that it takes approximately 25 years from the initial failure to its destruction by weathering and erosion on the undercliff. Processes at work on a slope therefore operate at *different scales*, some are large and occur infrequently, some are smaller and are continous. It is only through a long and detailed survey that our understanding of processes such as these can be increased. Unfortunately where landscape changes are slower few earth scientists have the patience (or live long enough!) to attempt to discover the *magnitude and frequency* of landscape-forming events.

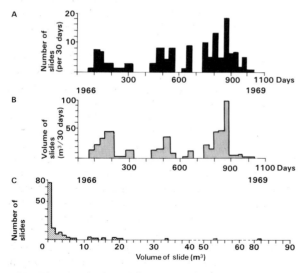

3.20 The magnitude and frequency of small scale slides on the undercliff

MASS MOVEMENT: THE PROCESSES

Mass movement has earlier been defined as a downslope movement of rock and soil in response to gravity. Classification on the basis of whether the movement involves creep, slides, flows or simply falls is possible. Nature is rarely that simple however, and movements occur in combination, as we have already seen in the case of the Dorset example.

Loose rock, stones and soil all have a tendency to move downslope. They will do so whenever the downslope force exceeds the resistance produced by friction and cohesion. When material moves downslope as a result of shear failure at the boundary of the moving mass the term *landslide* is applied. This may include a flowing movement as well as straightforward sliding. The *forces contributing to shearing stress* are the mass of the material (the rock, soil and the water contained within it) together with gravity.

The force of *gravity* acts constantly on all rocks and debris. Wherever there is a slope a proportion of the acceleration of gravity is directed parallel with the surface, in other words downslope. This downslope force increases with increasing slope angle, in fact as the sine of the angle of slope. For example, on a 30° slope the force acting on a particle is 50% of the gravitational force; on a 60° slope it is 87%. This simply means that the tendency to move is greater on steeper slopes. Between 0° and 45° there is a rapid increase in this tendency to move. Above 70° the slope virtually behaves as if it were perpendicular. Slopes exceeding 45° are relatively rare in nature; most of the surface consists of slopes less than 5° and on such slopes the signs of mass movement are not that dramatic in terms of the changes perceived in a human lifetime.

In addition to this gravity effect an additional dimension is the height of the slope, since pressure from material above is an element in the downslope force applied to material on the slope.

The resistance to these forces comes from friction and cohesion. In terms of mass movement we need to ask ourselves: *What factors lead to a decrease in shearing resistance?* Is it possible to divide these into three—materials, weathering changes and pore-water pressure changes. The role of water in all three is particularly interesting.

In terms of *materials* factors which lead to weakness include a host of bedrock features: in sedimentary rocks joints and bedding planes; in schists foliation, cleavage, brecciated zones and faults. Secondly, material may have a low internal cohesion—it doesn't hold together too well. Sands are an obvious example but some clays and organic matter exhibit the same properties. Thirdly, beds may decrease in shear strength if the water content increases, for example in clays, shale, mica, schist, talc and serpentine.

Weathering changes include in some cases the reduction in effective cohesion and shearing resistance. As a result of weathering, material may be more susceptible to movement on slope angles where the original material was stable.

The regolith, besides consisting of rock particles, minerals and organic matter, also contains voids or pores. At depth the pore spaces are filled with water. The upper surface of this zone is known as the water table, above which the pores are filled with air. After heavy precipitation, or as a result of human interference with drainage, the pores can be filled with water and the increase in *pore-water pressure* can lead to a reduction in shearing resistance.

In considering landslides—failures of material when the forces creating movement exceed the forces resisting it—we now need to examine the *factors leading to an increase in shear stress*.

These factors would include firstly a number of *transitory earth stresses* such as earthquakes, or in some situations heavy traffic. Increased *forces disturbing the slope* may also exist and these would include the accumulation of talus, snow and water. Again, although not significant at a global scale, man-made pressures might be important locally through the construction of embankments, dams or buildings.

Thirdly are a range of factors reflecting the *removal of underlying or lateral support*. This removal may be by natural erosional agents such as streams, waves or moving ice. It may be the result of weathering of weaker strata at the toe of the slope. In some cases the seepage of water through the regolith or rock may remove or wash out material resulting in less underlying and lateral support for the remaining material. Our old friend the bulldozer might also be responsible when making hillside cuts for roads or buildings.

Fourthly we return to *water*. As has been mentioned above its presence may lead to a reduction in shear strength. It will also increase the actual weight of the slope materials and thus increase stress. It is important to realise that water rarely acts as a simple lubricant, easing the downslope movement as it were; in some situations it can actually increase friction between particles.

You might now look back at the Dorset cliff system on page 58 and ask yourself what shear resistance reduction factors and shear stress increase factors might appear to have been operating.

Shear failure, producing landslides as defined above, is of course only one form of mass movement. *Creep*, although slower and less dramatic, is also important, especially if you consider that it is acting on all weathered slopes. Soil may expand and contract in response to changing temperatures and moisture contents. It may shrink when dry, swell when moist and swell when frozen. On a slope this movement is not just simply up and down. Particles will be lifted normal to the slope (at right angles to the ground surface) when expansion takes place. On contraction the particles return, but are pulled vertically back by gravity. A net movement downslope therefore occurs.

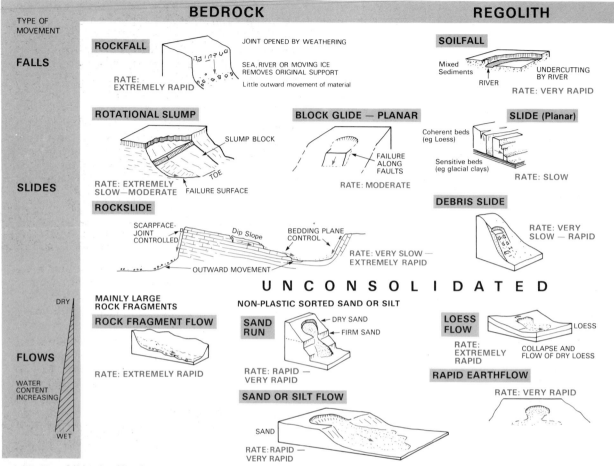

3.21   Landslide classification

It is useful to envisage not a single soil particle but rather the whole mat of soil very slowly zig-zagging downslope, held together by the vegetation roots. This movement ranges between 0·03 and 1·0 mm a year.

Finally there is the *flow* of material, a complex movement common in cohesive material such as clay. In theory flow movement is at a maximum at the surface and it decreases towards the base with increasing friction. In fact observations show that 'flows' frequently occur with a considerable degree of sliding over their base. As one of the characteristics of landsliding is that it occurs on a discreet boundary surface where failure has taken place such observations show how in nature it is sometimes difficult to neatly compartmentalise our classifications.

MASS MOVEMENT: THE FORMS

Figure 3.21 shows in a generalised way a classification of landslides based on whether the material involved is bedrock, regolith or unconsolidated, and whether the movement involved is predominantly falling, sliding or flowing.

Rockfall (top left) we have already examined in connection with scree slope development. Soilfall, with the unstable unit initially separated from the parent cliff by tension cracks prior to abrupt collapse, is similar. The slides and slumps (involving a rotational movement with a backward tilt on an upwardly concave failure plane or rupture surface) are familiar to us from the Dorset example.

(Regolith: rock fragments, weathered bedrock, weathered zone and soils)

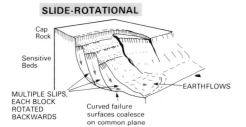

APPROXIMATE MOVEMENT RATES

3m/sec — EXTREMELY RAPID
0.3m/min — VERY RAPID
RAPID
1.5m/day
1.5m/month — MODERATE
1.5m/year — SLOW
0.3m/5years — VERY SLOW
EXTREMELY SLOW

SLIDE-ROTATIONAL

Cap Rock

Sensitive Beds

EARTHFLOWS

MULTIPLE SLIPS, EACH BLOCK ROTATED BACKWARDS

Curved failure surfaces coalesce on common plane

M A T E R I A L S

MIXED ROCK AND SOIL   MOSTLY PLASTIC

DEBRIS AVALANCHE   **SLOW EARTHFLOW**

RATE: VERY RAPID EXTREMELY RAPID

BEDROCK

REGOLITH

DEBRIS FLOW

RATE: VERY RAPID

3.22 The Madison slide, Montana, immediately after the earthquake with the dry river bed in foreground

very dependent upon lithological factors such as the jointing and bedding planes.

Earthflows and mudflows (bottom row) characteristically have a bowl-shaped source which leads into a narrow neck, or run, through which the material passes. Downslope this material is deposited as a series

On 17th August 1959 the sixth strongest earthquakes ever to affect the United States occurred in Montana. Close to the epicentre of the earthquake in the Madison River valley, a slope of schists and gneiss with slippery mica and clay in the weathered clefts and crevices was supported by a buttress of dolomite at its base. The quake cleanly broke the dolomite and the mountain face, 400 m high and 1,000 m long, failed and slid into the valley. 80,000,000 tons of material moved in less than a minute! The Madison River was dammed and a lake 60 m deep and 8 km long was formed before engineers cut a stabilised channel through the debris (Fig. 3.22). This is an example of rockslide (centre row) resulting from a transitory stress, the earthquake. The form of such rockslides is

3.23 Slumping of clay-rich shales, Coast Ranges, California

of overlapping lobes. They are frequently seasonal in behaviour, being more active in wet seasons when the ground water pressure is high.

Stiff, fissured clay may have a shearing resistance of 10–20 tons/m². If a cut is made—by a river for example—stresses are relaxed and cracks open up into which rain can pass, increasing pore-water pressure. The swelling of the clay also leads to the formation of clay blocks. As a result the shear resistance decreases—*i.e.* to 3 tons/m²—and as soon as it is equal to the average shear stress on a potential surface of sliding, the slope fails. In the London Clay, for example, the steepest natural slopes are 8° and are relatively stable, a slope of 18° fails after 50 years or so and a slope of 25° between 10 and 20 years. Figure 3.23 is an example of the landforms produced by this failure.

SLOPES

Everywhere you look the entire landscape is composed of slopes, even if they are so imperceptible that their inclination with the horizontal is almost undetectable to the naked eye. Because they exist everywhere they are of core concern to geomorphologists, and although there has been an advance in our understanding of slopes, Kirk Bryan's warning on 'not substituting words for knowledge and phrases for critical observation' still holds good. Slopes may be defined in two ways: either as the angle which any part of the earth's surface makes with the horizontal or in a wider sense as any geometric element of the earth's surface.

Slopes owe their origin to two groups of processes—endogenetic and exogenetic. *Endogenetic* processes are those reflecting conditions within the earth. Slopes can be produced by volcanism and by various types of earth movements such as faulting or folding.

Exogenetic processes on the other hand are those operating at or near the earth's surface. These include the weathering agencies and mass movements which have been introduced earlier in the chapter. They also include various processes of erosion and deposition by running water, moving ice, waves and wind. This group of erosional and depositional processes we will be meeting in the chapters which follow. The examination of slopes here can therefore be only a *partial* introduction to the topic. Without endogenetic processes the earth over long periods of time would be reduced to a surface close to sea level. Without exogenetic processes the landscape would consist of unmodified structural surfaces like fault scarps and volcanic slopes. It is the interaction between these two which produces the slopes in the landscape around us.

Lee Wilson in the *Encyclopedia of Geomorphology* listed some fundamental questions about slopes—

1. What is their actual form in plan and profile?
2. Do slopes decline through time or retreat parallel?
3. What is the importance of various processes on slope form and development?
4. What are the rates of operation of processes?
5. What are the variations of slope profiles with climate, lithology, structure, time, tectonism and relief?

At this level we can do little more than introduce them; some are, however, themes which recur later in the book and they are worth noting if you undertake further work and reading in this field.

Slope form and terminology

Initially we will focus our attention to geometrical terminology. Slopes can consist of elements which are concave upwards (the slope angle decreasing downslope), convex upwards (the slope angle increasing downslope) or straight (rectilinear). The term break of slope is applied to the point where the elements change. They can be combined, and this is shown on the left-hand side of Fig. 3.24.

The plan, as opposed to the profile of slopes, has been a neglected area of description. The right-hand section of Fig. 3.24 attempts to show four situations. Hack and Goodlet suggest convex radial-concave contour and concave radial-concave contour are the most common. Convex radial-convex contour seem to characterise many residual hills remaining after long periods of erosion.

King describes profiles distinctly different from the above. Figure 3.25 shows his profile consisting of a convex waxing slope, a rectilinear free face, a rectilinear debris slope and a slightly concave pediment. King claimed that this profile was universal—*i.e.* occurring in all environments—but it is best viewed as being limited to regions of well-stratified rock, as hillslopes on non-stratified rocks do not usually show this form.

Strahler in contrast divided erosional slopes into three types based upon the concept of angle of repose.

PROFILE

Convex

Concave

Straight
(Rectilinear)

Convex — Concave

Convex — Concave
with rectilinear

Complex

/ break of slope

PLAN AND PROFILE
FOUR COMBINATIONS OF CONCAVITY
AND CONVEXITY

CONVEX RADIAL-
CONCAVE
CONTOUR.

CONCAVE RADIAL-
CONCAVE
CONTOUR.

CONVEX RADIAL-
CONVEX
CONTOUR.

CONCAVE RADIAL-
CONVEX
CONTOUR.

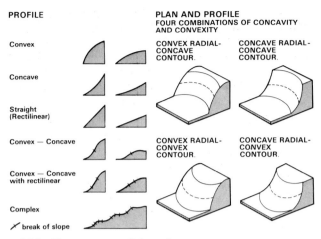

3.24   The geometry of slope forms

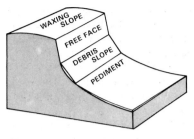

THE FOUR ELEMENTS IN A FULLY
DEVELOPED HILLSLOPE PROFILE
ACCORDING TO L. KING

3.25   The four elements of a fully developed hillslope profile

Slopes lying at the angle of repose of the material were known as repose slopes. Steeper slopes were known as high cohesion slopes—for example, those developed on resistant bedrock or dry compacted clay. Gentler slopes, subjected to creep and to the various wash processes to be examined in the next chapter, were known as reduced slopes.

There are a number of other classification systems but it should be emphasised that slope form can only

accurately be known from detailed survey; the eye can deceive and the contour interval used on maps may as easily conceal differences as display them.

Parallel retreat or slope decline

The roots of this hoary old argument go back over 100 years. In 1785 Buffon argued that slopes decline through time, in 1886 Fisher argued that slope retreat is parallel for vertical cliffs. The nature of slope change through time has emerged as a recurrent theme in geomorphology ever since. During this century W. M. Davis envisaged slopes declining with time in humid climates, whilst Penck and King (amongst others) have considered parallel retreat and slope replacement to be the primary ways in which slopes change. These views are represented in Fig. 3.26. We will return to this issue later in the book. In summary, it is probably sufficient to say that there is no single theory of hillslope evolution applicable to all environments which have different structures and varying combinations of processes operating within them.

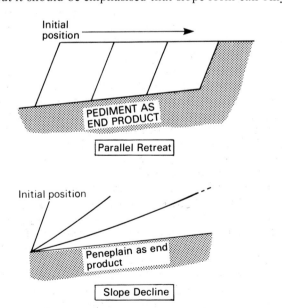

3.26   Two views of slope change

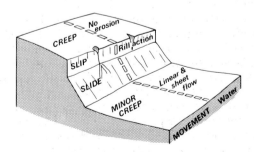

3.27   Slopes and processes

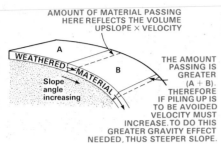

3.28   Hilltop convexity and material movement

The importance of various processes on slope form and development

We have in fact already explored part of the dimensions of this question in connection with the two case studies of screes and the Dorset cliff. The processes we are concerned with are those of mass movement and weathering and in addition the role of surface runoff and ground water conditions. The latter two are best considered in connection with the drainage basin as part of the next chapter. Figure 3.27 indicates how King related processes to the form of slopes.

An interesting example of how reasoning about process-form relationships takes place is provided by G. K. Gilbert's deduction that slope angle must increase downslope to transmit material from above. This is shown in Fig. 3.28. You might think of a number of additional questions, *i.e.* why the slope becomes convex in the first place and a range of questions revolving around the regolith—what is its thickness, how fast does it move, etc. These point us in the direction of the need for measurement of processes affecting the hillslope.

The rates of operation of slope processes

This is a fundamental problem and the achilles heel of many earlier slope theories. Processes vary in space (from one part of the slope to another and from one

Process	Volume (m³)	Tons/km²	Average Movement (m)	Tons moved per vertical metre
Rockfalls	50	8.7	90 — 225	19,565
Avalanches	88	15.5	100 — 200	21,850
Earthslides	580	69.4	0.5 — 600	96,375
Talus creep	300,000	—	0.01	2,700
Solifluction	550,000	—	0.02	5,300
In solution (running water)	150	26	700	136,500

3.29   The processes in Karkevagge

PROCESS	LINEAR RATES (cm/p.a.) 10° slope	VOLUMETRIC RATES (cm³/cm/p.a.) 10° slope
SOIL CREEP (vegetated)	<1	2.0 - 5.0
TERRACETTES (vegetated)	5-10	20
SOLIFLUCTION (unvegetated)	5-20 (very variable up to 100 cm/day)	50
RAINSPLASH (unvegetated)		
20 mm stones	0.2	
2 mm stones	20	200
0.2 mm stones	150	
UNGULLIED (unvegetated) SURFACE WASH		<1000
LANDSLIDES	100 (very variable)	
ROCKFALL/SCREES (25-40° slope)		1-5

3.30   Typical hillslope movement rates

area to another) and in time. The Dorset example has shown that some processes are virtually continuous whilst others occur spasmodically.

What is the relative significance of the various slope processes? Figure 3.29 shows the results of a classic study by Rapp in a high-latitude mountain environment. Look carefully at the various column headings for materials and distances involved. Talus creep and solifluction involve large volumes of material yet the movement is small and hence the tons moved per vertical metre is modest. Solution on the other hand involves a smaller volume of material during the time period but it is moved over greater distances.

Comparable data for other environments is unfortunately not available. Figure 3.30 shows some typical movement rates on a 10° slope. The solifluction process mentioned occurs in periglacial environments where thawed surface soil flows over a permanently frozen subsoil. Terracette movement may also need some explanation. Terracettes are narrow flat-topped steps along moderate to steep grass slopes. They often form anastomosing lines on the hillside and in England

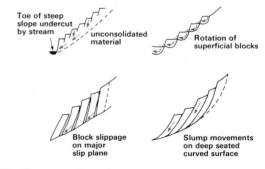

3.31   Terracette formation

3.32 Mesas and buttes, Utah. 250 m cliffs of Permian sandstone are capped by more resistant layers of the Moenkopi formation

3.34 Chalk slopes with terracettes

are frequently known as 'sheep runs'. They may represent miniature rotational slips or other processes shown in Fig. 3.31. As the actual surface shapes may be the same they are another example of equifinality of form, already mentioned in the case of surface depressions. The figures for rainsplash and wash are included for comparative purposes as the discussion of these processes occurs in Chapter 4.

Variations of slope profiles with lithology, structure, tectonism, climate, time and relief

Isolating the individual influence of this range of factors is difficult. They interact with each other and with slope processes to produce a variety of slope forms. The effect of geological factors can be seen in particularly striking form in arid environments, where the lack of a vegetation cover exposes the effect of rock structure and character (Fig. 3.32). Soil and vegetation mats often mask such variations in slope form. Differing susceptibilities to weathering and mass movement do produce detectable variations in slope angles (Fig. 3.33) and slope forms (Fig. 3.34).

The inclination of sedimentary rocks of differing character results in a range of landforms—the flat-topped, steep-sided mesa on horizontal layered rocks (Fig. 3.32) and *cuestas* (escarpments), 'flat irons' and 'hog backs' on inclined strata. Figure 3.35 is a geological and relief section in the Rocky Mountains. The detail of the geological annotation (in blue) should be

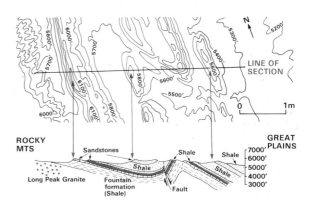

3.35 Geology and relief

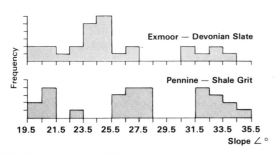

3.33 Slope angles and lithology

3.36 A fault scarp in Utah

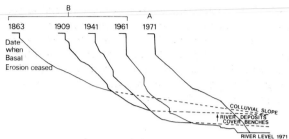

3.37   Sequential slope profiles of a river cliff

related to the relief forms. Finally tectonism (such as the faulting of an area) can produce steep-sided straight initial slopes which are modified by weathering and mass movement. Figure 3.36 is an example of such a *fault scarp* undergoing dissection by the action of weathering, mass movement and running water.

There are only a few situations where an assessment of slope profile changes over time can be undertaken—Fig. 3.37 shows a series of river-cliff profiles in Louisiana. A indicates the situation where the river was still undercutting the cliff and continually removing materials. B profiles represent conditions where the river has moved away from the cliff as a result of the normal migration of the channel in the meandering course. What appears to be happening to the shape of the slope and the proportions of the various slope elements as the river undercutting ceases?

Slope development through time has been mentioned above and it is a theme which will be returned to in the context of the drainage basin.

CONCLUSION

The treatment in this chapter partially links material introduced in Chapters 1 and 2. Endogenetic processes within the lithosphere produce initial slopes composed of a variety of rock types. Exogenetic processes, reflecting an interaction between the atmosphere and lithosphere, mould and modify these initial surfaces. In the context of this chapter these processes are primarily those of decomposition, disintegration and mass movement. Agents of erosion such as running water, moving ice and wind are equally significant factors in the origin of landscape and these form the focus of the chapters to follow. Finally this chapter has also introduced a number of case studies, which have

attempted to outline various approaches to geomorphological investigation—theoretical, laboratory and field—and to introduce the ideas of systems analysis.

Review Exercises

1. *Slope Survey Project.* The pantometer (Fig. 3.38A) will, in one traverse up a slope (at right angles to the contours), produce data which can be used to draw either a scale profile or a slope-frequency histogram. The latter allows the data to be statistically described and different slope profiles to be compared.

On the basis of an understanding of slopes, three hypotheses could be erected: (i) that lithology influences slope (this can be more precisely stated in the light of the geology of the study areas chosen); (ii) that basal removal produces a steeper slope (see Fig. 3.37); and (iii) that exposure, or aspect, influences slope. The latter reflects differences in dampness, temperature, etc., which may influence weathering and mass movement processes operating at the present or during the Pleistocene. (Solifluction is discussed in Chapter 6.)

The next step is to plan the research design carefully to ensure that procedures are consistent and that data is obtained to test the hypotheses. It is often useful to try and simulate laboratory conditions and to try and exclude the effect of other variables by holding them constant. In the case of the three hypotheses mentioned it is obviously important to exclude major climatic contrasts in the study areas. Land use may affect slope processes, too, so sites could be restricted to pastureland (this makes the survey easier and more accurate, too!). Finally one might select slopes only from valleys of equal magnitude, *i.e.* unbranched first-order valleys (Fig. 3.38B). What is being treated as 'noise' here could of course be equally valid aspects to investigate.

The project falls into three parts. Firstly, the where, why and what aspects introduced above. Secondly, the actual data collection in the field (an example of a field record is shown in Fig. 3.38C). Finally there is the collation and analysis of the observations. Individual slope histograms may be drawn and described (mean, mode, etc.). They may be drawn together for collation and comparison, lithology X compared with lithology Y, basal removal against non-basal removal, northeast against south-west-facing slopes and so on. There may be variations of slopes within these broad categories, basal removal may perhaps produce a bimodal distribution. Is it related to the two rock types? This kind of project may very well throw up more questions than you started with! If this happens try and think of other investigations which you could undertake.

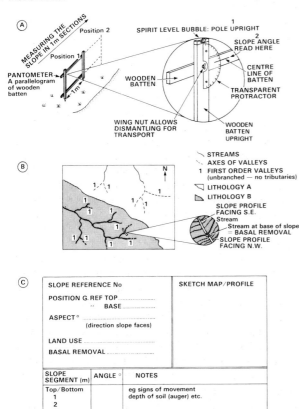

3.38 Slope surveying

2. *Slope modelling.* On graph paper draw a 100 mm vertical line, representing a fault scarp. This weathers back evenly, producing an equal volume of talus lying at 35° at its base (Fig. 3.39). In the next time period the cliff weathers back 10 mm, but only from the exposed face. Measure the volume involved and spread this evenly over the talus. Repeat the process until there is no longer a cliff contributing debris. What do you notice about the proportion of the slope elements through time and the shape of the bedrock surface?

What happens if we (i) assume half the debris has been removed (*i.e.* by a stream), (ii) assume that the debris slope itself weathers to produce material at its base with an angle of repose of 10° (iii) vary the angle of repose of the debris. (You could experiment with sand, gravel, dried peas, etc., to discover how the angle varies with the grain size/texture.) Describe the profile changes through time. Does this simulation in any way represent real world conditions?

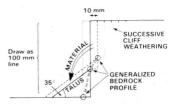

3.39 Slope modelling

Review Questions

1. Describe the processes operating in Figs. 3.5, 3.6, 3.7 and 3.15.
2. What are the processes of rock disintegration and decomposition?
3. Describe and evaluate the role of water in mass movement.
4. Look at Fig. 3.18H. Discuss, with your reasons, what you think may happen next.
5. Why and in what way do slopes differ in form?
6. Outline some of the ways in which geomorphologists have studied weathering and slope processes.

Further Reading

Bloom, A. L., 1978, *Geomorphology*, Prentice Hall.

Brown, E. H., & Waters, R. S., 1974, *Progress in Geomorphology*, Institute of British Geographers Special Publication **7**.

Brunsden, D. (Ed.), 1971, *Slopes, Form and Process*, Institute of British Geographers Special Publication **3**.

Brunsden, D., & Doornkamp, J., 1973, *The Unquiet Landscape*, David & Charles.

Carson, M. A., & Kirkby, M. J., 1972, *Hillslope Form and Process*, Cambridge University Press.

Chorley, R. J., & Kennedy, Be, 1971, *Physical Geography: a Systems Approach*, Prentice-Hall.

Gardner, J., 1977, *Physical Geography*, Harper & Row.

Young, A., 1972, *Slopes*, Longman.

4 Water on the Land

4.1 The moon's surface; landforms produced without water

4.2 The earth; the effects of water action. A Landsat image of the Himalayas and the Ganges plain

THE APPROACH OF THE CHAPTER

The two images alongside show more dramatically than mere words the effects of water in shaping landscape. The surface of the moon, pockmarked with impact craters, represents an environment whose detail is being shaped by processes operating without an atmosphere and without the presence of running water. The landscapes of earth, on the other hand, are being moulded in environments where water is continually cycled between the atmosphere, the land and the oceans.

This chapter examines the role of water in the landscape. It begins with a study of the pathways of water from precipitation to its ultimate arrival in the sea, the basin hydrological cycle. The drainage basin will be considered as a system, with energy, water and rock material inputs, outputs and stores. Basins of different sizes and in a range of environments will be examined, together with the landforms produced by the erosional, transportational and depositional work of rivers.

Finally the chapter will also concern itself with the theme of water as a resource for mankind; for quenching his thirst, for irrigating his croplands, for use in industry, in transport, in power generation and for recreation. To do this we will have to integrate not only facets of the physical environment, but also the manner in which individuals and society perceive of water resources and how they decide to manage them.

THE BASIN HYDROLOGICAL CYCLE

The concept of the hydrological cycle has already been introduced in Chapter 2. Water vapour in the atmosphere is precipitated onto the land after condensation occurs. Once on the land water may be evaporated back to the atmosphere from the ground surface itself or by way of the root-leaf transfer of transpiration.

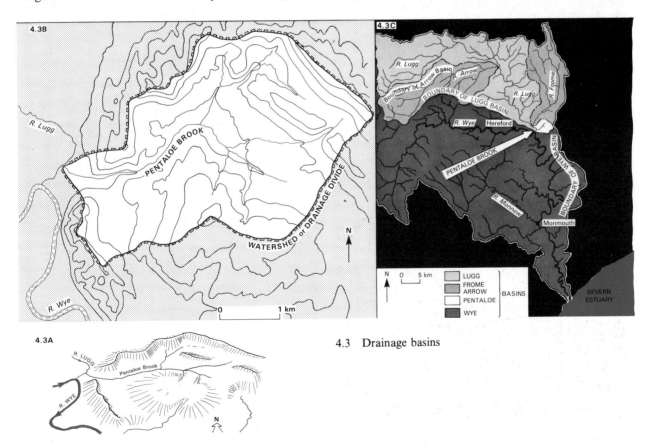

4.3   Drainage basins

The remainder can sink into, and beneath, the surface and is either moved slowly to the sea or it passes much more quickly by way of surface stream-flow.

The treatment in Chapter 2 was in terms of the global cycling of water molecules. The discussion here is in the context of the *drainage basin*. A drainage basin is an area of the earth's surface bounded by a watershed or stream divide. It is therefore the catchment area drained by a particular stream. Figure 4.3A and B show this by means of a perspective sketch and relief map.

Individual streams and their catchment areas join together to form larger basins as the individual streams combine to form tributaries of the larger streams and rivers. The surface of the land is therefore covered by a *nested heirarchy* of drainage basins. This is shown in Fig. 4.3C where the Pentaloe Brook catchment can be seen to be part of the Lugg basin, which in turn forms part of the Wye drainage basin.

The drainage basin as a finite area of the earth's surface can be viewed as a *system*, a set of objects (stream channels, slopes and so on) together with the relationships between them. The basin has a boundary, its watershed, and it is maintained by a constant supply and removal of energy. It is therefore viewed as an *open system* in which solar energy provides the motive force and within which precipitation is converted to runoff by a range of processes which move and store water.

The drainage basin is shown diagramatically and as a system in Fig. 4.4. In the section which follows we will trace the routes followed by water after its input as precipitation over the basin.

THE PATHWAYS OF WATER

Precipitation

Precipitation, of all the climatic elements, is probably the one with the longest history of measurement. In spite of this there are two fundamental queries: just how accurate are the gauges and how representative of an area are the figures obtained?

Gauges are of two types—the non-recording and the autographic or recording gauge. A non-recording gauge is simply a storage device from which the rainfall between observer visits is measured. Figure 4.5A shows the standard Meteorological Office gauge. For many applications, such as storm-drainage design, the actual intensity of rainfall and its duration is required, not merely the six- or twenty-four-hour total. The recording, or autographic, gauge shown in Fig. 4.5B measures this dimension of rainfall.

Gauges such as these cause turbulence and eddying. Raindrops may be swept clear of the opening and the gauge under-record the rainfall. Gauges mounted at ground level, surrounded by an open wire grid to

4.4 Components of the basin hydrological cycle

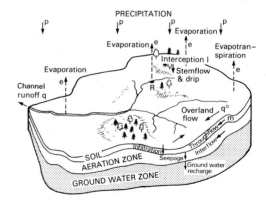

remove the problems of splash, indicate that the conventional gauge may underestimate rainfall by as much as 20%.

You might have noticed the deliberate use above of the term rainfall. In many areas precipitation occurs as snow which has its own peculiar measuring problems. Light snowfall can be melted in the conventional gauge. For heavier falls a non-recording gauge, but with a much bigger store (1 cm of snow is equivalent to 1 mm of water) might be required together with a shield to reduce the effects of drifting and turbulence. Drifting of course is another problem in itself—snow has the annoying habit of moving in the wind after it has fallen! Recording snow gauges work on the principal of weighing the snowfall, *i.e.* expressing it as a water equivalent. This could involve collecting the snow in an antifreeze solution or using a snowpillow to continuously weigh the snowpack (Fig. 4.5).

When you add to these problems the fact that gauges have to be made and read by fallible humans you might become cautious about precipitation figures. The writer remembers a recording gauge costing as much as a small car which expired after a few

weeks of a Labrador winter. There is also the apocryphal story of a Canadian weather station which suddenly began to record incredibly low snowfalls—the 'observers' had moved the gauge to a position just outside the door so that they didn't have to spend too long outside in forty degrees of frost! Consistency in instrument design, siting and recording offset these problems.

The total surface area of all the rain gauges in the United Kingdom is 0·0013 km^2. Compared to the total land area this is a minute proportion. It emphasises the fact that gauges are *point samples* of an element which occurs over the whole surface. Two of the methods used to estimate precipitation over a basin area are shown in Fig. 4.6.

After these cautionary tales about measurement it is now time to look at the character of this precipitation input into the basin cycle. In particular two points concern us: *its distribution in time and in space*. You are familiar with the way in which these are conventionally represented, *i.e.* by precipitation maps with isohyets, and graphs showing the monthly totals, both usually based on average precipitation over a sus-

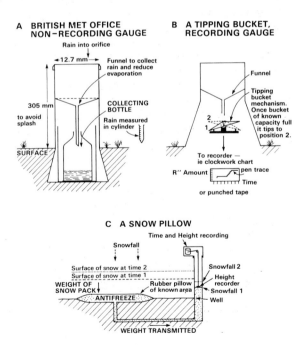

4.5   Measuring precipitation

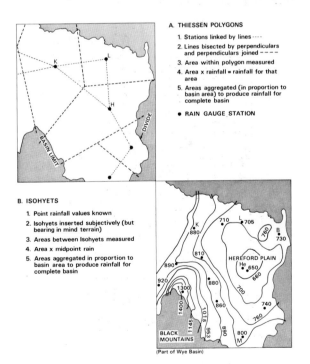

4.6   Determining rainfall over a basin

tained period of observation such as 30 years. The magnitude and character of some seasonal differences will be taken up later in the chapter.

Precipitation may vary across an area. Figure 4.6B shows this in the case of the central part of the Wye basin where there is a close correlation between precipitation and relief. The annual 1,145 mm isohyet to the west of the centre of the map corresponds to the 600 m high mass of the Black Mountains and the 660 mm isohyet corresponds to the lower Hereford Plain. The relief effect, where forced uplift of air over mountains augments frontal precipitation, may be repeated at smaller scales than that shown in Fig. 4.6— over individual hills and ridges, for example.

At even *smaller scales* vegetation may influence the distribution of precipitation. Figure 4.7 shows this in the case of snowfall over a tiny area of Labrador–Ungava in Canada. This area has gentle relief, consisting of a series of north-south trending ridges about 10 m high. The better-drained ridge tops are covered in woodland whilst bog and muskeg occupy the lower sites (Fig. 4.7A). The prevailing wind is from the west. The pattern of late winter snowdepth, and the snow's water equivalent, are shown in Fig. 4.7B. How would you describe its relationship with vegetation? Under-

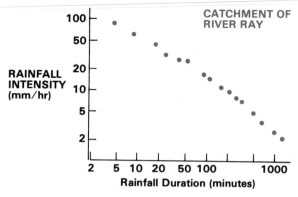

4.8   Rainfall intensity and duration

standing such variations is important when assessing the reliability of an individual station's snowfall records as indicators for a wider area.

Two other dimensions of precipitation relevant to the basin hydrological cycle need mention. These are shown in Figs. 4.8 and 4.9. The former shows the rainfall *intensity* and its *duration* for 16 storms over the Ray catchment. Notice that the highest intensity storms are of short duration. This kind of relationship is universal, although the 'slope' of the relationship varies from region to region. If you think back to the origins of precipitation in Chapter 2 you could assume that the high intensity short-duration storms are convectional in origin, the longer period lower intensity

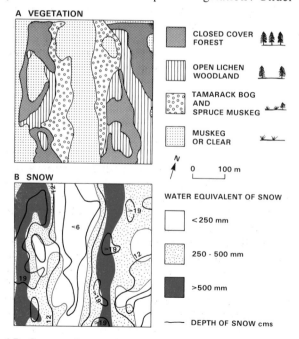

4.7   Snow and vegetation

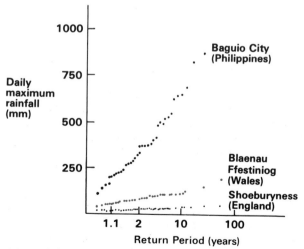

4.9   Rainfall frequency

events reflecting the passage of frontal rainfall. Convectional storms, besides being of short duration, often cover only a limited area—thus part of the catchment could experience heavy precipitation whilst other parts are dry.

Figure 4.9 shows the *return periods* of maximum rainfall for a period of years. The return period is produced from a list of maximum 24-hour rainfalls for a period of years. These are ranked from highest to lowest and the return period calculated $(n+1) \div m$ (where n = number of years of record and m = rank). Figure 4.9 shows that higher rainfalls occur at longer return periods, but there are variations between the three stations in the magnitude of the daily rainfall for any specific return period. The actual slope of the relationship is also different. Can you think of the climatological reasons for these variations?

Interception

If you look back at Fig. 4.4 and follow the precipitation input from the top centre of the diagram you will notice that part of the precipitation may fall directly onto the stream channel. It has therefore entered the channel store (water in the stream itself) and exits from the basin as channel runoff. Since streams rarely occupy more than a few percent of the basin's area, most of the precipitation enters the *vegetation subsystem*.

It is useful to imagine rain (or snow) falling onto a plant cover. For a while after the storm has started some of the raindrops may be held up on the leaves or branches of the plant. This is known as *interception*.

Interception varies with the duration of the storm and with the character of the vegetation. If the storm lasts a short time a considerable proportion of the rainfall, perhaps all the first millimetres, remains caught on the leaves and branches (so called interception storage). In a longer storm these myriad small reservoirs overflow and the rainfall drips to the surface or flows down the stems. As the duration of the storm increases so too does the proportion of rainfall reaching the surface beneath the plants. Put another way the proportion intercepted falls as the storm continues. In addition, in the case of a short storm, once the rain has stopped falling the stored water may be directly evaporated back to the atmosphere and play no further part in the basin cycle. The same occurs after the end of any period of rainfall.

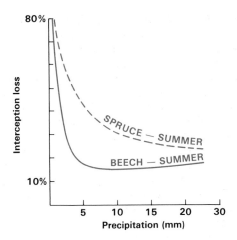

4.10 Interception of precipitation by spruce and beech

The character of the vegetation itself is also significant. The height, density, proportion of surface covered and the seasonal changes in foliage all produce variations in the proportion of precipitation intercepted. Some examples of these are shown in Fig. 4.10 which also indicates the effect of precipitation amount on the proportion intercepted.

When examining the routing of water within the basin cycle the effect of the vegetation cover therefore has to be considered. It varies across the basin and also through time, seasonally with the plant's life cycle and over longer periods with land use changes.

Infiltration

If you return again to Fig. 4.4 you will see that we have traced the precipitation as far as the ground surface. Some has fallen directly onto the surface and some has arrived as stemflow and drip. We now arrive at a crucial 'valve' in the system—the capacity of the soil to transmit the water which falls on it. The rate at which water can pass into the soil is known as its *infiltration capacity*, usually expressed as mm/hr. If the water falls at a rate greater than the infiltration rate it will be held back on the ground surface. It may form puddles—surface storage—or it may flow off downslope to the stream channel, a process known as overland flow. After a storm, water from the surface store may be evaporated, moving off to the right of Fig. 4.4.

Water is actually drawn into the surface by two forces—gravity and capillary attraction. Capillary water is that which is held as a thin molecular film around the soil particles. As rainfall continues and the spaces between the soil particles becomes filled with water capillary forces decrease and gravity pulls the water further down into the soil. Thus if infiltration is measured during a storm it tends to start at a higher rate and then fall to a relatively steady rate.

Infiltration is obviously influenced by how much moisture is in the soil before a particular storm. This will reflect rainfall over a period preceding the storm, and also the topography and configuration of the slopes in the basin. The significance of this latter point will be explained later.

In addition to the moisture content of the soil another factor is the existence of 'big holes and little holes' in the soil, to use Darrel Weyman's graphic phrase. The 'little holes' are determined by the texture of the soil, which in turn is influenced by the mineralogical qualities of the parent material (*i.e.* bedrock or transported material) and the weathering it has experienced. Screes, which were mentioned in the last chapter, consist of large irregularly shaped rock debris and their considerable void space can transmit massive amounts of water. At the more typical size range of soil particles, sands (>0.2 mm diameter) can transmit 200 mm/hr. The finest particles, clays (<0.002 mm diameter) on the other hand have very low infiltration capacities, often less than 5 mm/hr. The 'big holes' are the larger spaces between the elements of the soil structures (the groupings of the individual particles) together with root passages.

The role of *vegetation* is considered to be a very important influence on infiltration. The presence of plants promotes a thicker soil, and a freer draining texture. The vegetation also breaks the impact of the falling raindrops which can compact the surface and dislodge small particles to clog the water passageways. For example, infiltration rates on the same soil can range from 57 mm/hr on a permanent pasture, through to 13 mm/hr on heavily grazed pasture and to as little as 6 mm/hr on bare crusted ground where this clogging process is most effective.

The measurement of infiltration rates is discussed in the readings at the end of the chapter.

Throughflow

At this point we have now traced the water into the soil and its vertical movement downwards. Most soils are more compact as depth beneath the surface increases. The soil's permeability—its capacity to transmit water—therefore decreases with depth, especially when there are clay layers or hard pans present within the soil. When this happens the water is deflected laterally downslope within the soil, especially through the larger spaces. This downslope movement within the soil is known as *throughflow*. It is shown in Fig. 4.4 where the infiltrating water is routed to the left of the diagram towards the stream channel.

Within this subsystem there is also a potential loss of water. Evaporation and drying at the surface draws water upwards by capillary attraction. The process of plant transpiration also draws water vertically upwards from the soil moisture store, from the root zone to the stomata on the leaves. These two processes combined are known as *evapotranspiration*.

Throughflow occurs particularly where vegetation, animal burrows and drying produce cracks—'big holes'—in the soil. When this is combined with decreased permeability in the soil, lateral downslope movements of water are encouraged. This occurs in an interconnecting series of passages, up to several centimetres in diameter, known as *pipes*. Their existence and significance in transmitting water, and of course in transmitting weathered material downslope, has only been realised since the 1960's. They require a soil strong enough to keep them open between periods of rainfall; cohesive peats are an example in upland catchments in the British Isles. Water transferred in this way—*i.e.* as throughflow—passes into the river channel at the base of the slope, either as slow oozings of moisture or directly from the subsurface pipe network.

Deeper and slower transfers

Beneath the soil some water may penetrate into the underlying rock, percolating vertically under the force of gravity. All rocks show some porosity (the volume of voids as a percentage of the bulk volume of material) and permeability. The latter, *permeability*, is a rock's capacity to transmit water, in other words its hydraulic conductivity. Across the whole range of rocks permeability varies by 10^9. Fine-grained igneous rocks and fine-grained sediments such as clays show the lowest permeabilities. Coarse-grained sedimentary rocks show the highest permeabilities. Permeability depends on grain size, shape, variation and packing and on the existence of joints and bedding planes in the rock. Limestones, for example, often show a low matrix permeability but a high secondary permeability. The actual chunks of solid limestone have a limited capacity to transmit water yet the joints and bedding planes, especially when solutionally enlarged, give a limestone area as a whole a permeable characteristic.

The percolating water eventually reaches a zone of saturation, where all the voids are water filled, the so-called ground water zone. This is shown as the bottom store on Fig. 4.4. The upper surface of this zone is known as the *water table*. The water table can vary seasonally of course with the amount of precipitation falling on the catchment and the proportion of this which reaches the ground water store. In which season would you expect most water to reach this zone?

The water table usually forms a subdued image of the surface topography. This is shown in Fig. 4.11. In less permeable rocks the water table has steeper slopes, reflecting the increased resistance to flow in impermeable rocks. In this case its surface more closely mirrors the surface topography.

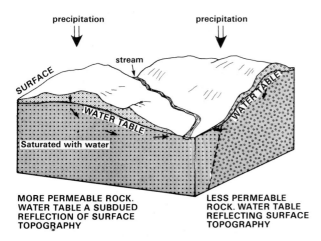

MORE PERMEABLE ROCK. WATER TABLE A SUBDUED REFLECTION OF SURFACE TOPOGRAPHY

LESS PERMEABLE ROCK. WATER TABLE REFLECTING SURFACE TOPOGRAPHY

4.11   Ground water and water tables

The idea of a single water table surface may in many situations be too great a simplification. Figure 4.12 shows the situation on the Carboniferous Limestone of the northern flank of the Mendip Hills (see also the map on p. 53). Water which sinks into the limestone at A has been traced to emerge at 2, a distance of 1,000 m, after 42–46 hours. Water sinking at A has also

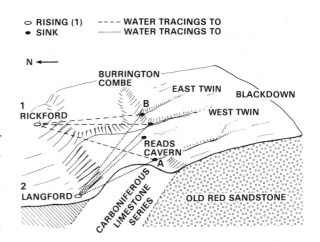

4.12   Water tracings on the north flank of the Mendips

been traced to rising 1, a distance of 2,300 m which it travels in 4 hours. From sink B water has also been traced to 1 and 2, and a number of other routes are also shown in Fig. 4.12. The flow lines between sink and rising cross without water mixing and from some sinks they proceed to separate risings. Water must therefore travel in distinct conduits in the massive crystalline Carboniferous Limestone.

The situation shown in Fig. 4.11 views the rock as an isotropic medium—*i.e.* of uniform properties in all directions. Within the phreatic zone, beneath the water table, water moves along the steepest hydrological gradient towards the springs and seepages. Chalk, with its dense network of micro-fissures and slow laminar flow of water through them, was believed to resemble this model. However, more recent work has detected zones of high transmissivity in chalk, similar to those in limestones.

Our knowledge of the deeper and slower transfers within the basin cycle is naturally limited. Broadly speaking it is only in limestone areas that any extensive water tracing (by means of dyes or spores) has been undertaken. An area with a dense network of wells for public water supply can also provide information on variations in the depth of the water table and its seasonal changes.

In conclusion, the significance of ground water in our discussion of the basin cycle stems from two facts. Firstly the water moves more slowly than the transfers mentioned earlier. Secondly the actual quantity of water in the ground water store is huge compared to the other stores already mentioned. Combining these two facts means that ground water can maintain river flows over long periods of time between discrete precipitation events. Ground water is therefore the reason for rivers continuing to flow during droughts and dry seasons.

The rates of water movement

In the sections above, the input of water into the basin hydrological cycle has been traced through the various stores and transfers. Water moves at different rates through the system. These velocities are shown in Fig. 4.13 which should be looked at in conjunction with Fig. 4.4.

	MEDIUM OF FLOW	VELOCITY m/hr
Surface	OVERLAND FLOW	50-500
Soil	PIPEFLOW MATRIX THROUGHFLOW	50-500 0.005-0.3
Deeper Transfers	GROUND WATER FLOW Jointed Limestone Sandstones Shales	10-500 0.001-10 0.00000001-1
The stream	OPEN CHANNEL	300-10,000

4.13   Flow estimates for hydrologic processes

WATER FLOW OVER TIME

So far only the pathways of water have been explored. It is now time to look at how the water input,

transfer and storage combine to produce the output from the basin. This will be done by looking at basin behaviour over the period of a year and then by examining shorter time periods and the question of floods.

Water balance

The water balance of a catchment is the relationship between input and output: $P = Q + E \pm \Delta s$. Precipitation equals runoff plus evapotranspiration, plus or minus any changes in storage.

The problems of measuring the amount of precipitation input have been mentioned above. Evapotranspiration losses are even more difficult to assess. Within the basin water is found at the surface on vegetation, in the various surface stores and in the stream itself. The first two tend to be only temporary, occurring most frequently during storms when direct evaporation is at its lowest. Transpiration, the vertical movement of water in plants, is therefore the most significant way in which water is returned to the atmosphere. A broad-leaved woodland in summer can in fact transpire as much water per square metre as could be evaporated from an open water surface.

Runoff, the actual discharge of water from the basin, can be measured in a number of ways. Some of these are shown in Fig. 4.14 and are mentioned in the further readings at the end of the chapter.

In view of the problems involved in trying to measure storage changes water balance work is best

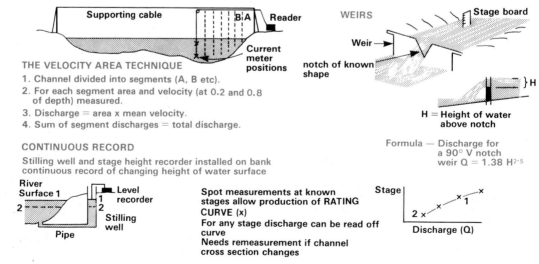

4.14   Measuring discharge

Station, Location & Area		O	N	D	J	F	M	A	M	J	J	A	S	Year
Haweswater Beck, Lake District	Rainfall	182	168	295	229	364	215	191	234	174	84	234	309	2679
(NY 515161) 34 km²	Runoff	169	132	367	225	343	186	197	200	122	41	185	284	2451
R. Wye, Rhyader, Mid-Wales	Rainfall	77	143	451	95	224	106	120	132	139	113	119	100	1819
(SN 969676) 167 km²	Runoff	59	82	384	102	165	89	94	77	71	39	62	50	1272
R. Adur, Sussex	Rainfall	13	102	152	44	124	20	111	65	66	76	80	39	892
(TQ 178197) 109 km²	Runoff	2	35	97	42	82	7	56	29	3	1	6	3	362
R. Glem, Suffolk	Rainfall	16	56	95	27	45	14	50	39	62	91	60	20	575
(TL 846672) 87 km²	Runoff	7	13	61	22	31	10	16	7	4	3	4	3	182

4.15   Water balances 1965–6

started and finished at a time of minimum storage when errors have the least significance. In the British Isles this is autumn, so the 'water year' runs from 1st October to 30th September. Figure 4.15 shows the water balance in four catchments for the year 1965–6. The rainfall figure is for the whole catchment above the gauging station. The output from the basin is expressed as runoff (mm)—*i.e.* the volume of water passing the gauge during the month is represented as a thickness of water from the whole area. This makes it comparable with the rainfall figure whatever the basin size.

If you look at these figures in all four basins rainfall exceeds runoff for the year, but the proportion lost varies. You might check the location of these basins and think of reasons for the variations in rainfall and runoff. In connection with the losses (the difference

between rainfall received and runoff), temperatures and plant growth are particularly relevant. In the two western and upland catchments in some winter months runoff actually exceeds rainfall! What processes within their basin cycles might account for this?

In contrast to these Fig. 4.16A shows the water balance for a subarctic catchment. Rainfall, snowfall, stream runoff and evapotranspiration measurements have been made for each month. The month by month pattern of precipitation is typical of an eastern Canadian subarctic climate. To what extent is it reflected in the monthly runoff figures? Using Fig. 4.4 to organise your ideas you should note that the runoff pattern reflects the existence of two large stores which change seasonally—the winter snowpack, a surface store lasting from autumn until May (Fig. 4.7), and the 24% of the basin occupied by lakes (Fig. 4.16B).

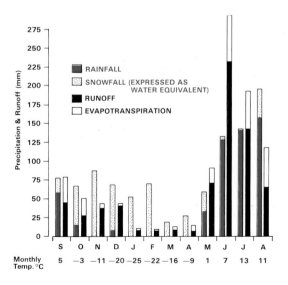

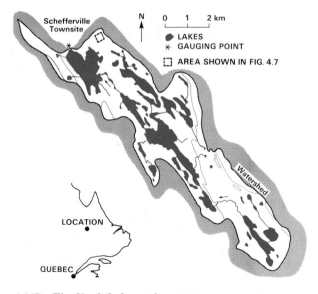

4.16A   Monthly water balance for a small subarctic catchment, Knob Lake, Quebec, 1964–5

4.16B   The Knob Lake catchment

The runoff pattern through the year is known as the *river regime*. The five examples here relate only to one year. If you have looked at the runoff and rainfall figures and pondered about the processes which they reflect you will have realised that the output of water from a drainage basin reflects a host of interacting factors relating to the climatic, geological and biogeographical (vegetation and soils) characteristics of the basin. Later in the chapter, in the context of environmental contrasts and resource management, we will be looking at river regimes over a period of years.

The storm or flood hydrograph

After the onset of a storm over a basin there is often a steep rise in runoff, the flood peak, which is followed after the rain stops by a gently falling recession limb as the discharge drops (Fig. 4.17A).

In view of the relatively short length of time between the start of rainfall and the peak (the lag) the water must have reached the channel by way of various *quick flow* processes. The slower and deeper transfers, the *delayed flow*, produces the *base flow* for the stream. Base flow shows much less pronounced variations over short time periods (*i.e.* hours) but obviously reflects seasonal variations in precipitation, evapotranspiration and vegetation.

During heavy rainfall water may be seen on the surface, so-called infiltration excess overland flow. If a histogram of storm rainfall is superimposed on the infiltration curve for a catchment the volume of water left on the ground surface can be defined. This is the Horton *overland flow model* of the storm hydrograph. Overland flow is envisaged as occurring simultaneously over the whole catchment once the rainfall intensity has exceeded the infiltration capacity. Overland flow moves at velocities of up to 500 m/hr which would mean that areas within 500 m of the stream channel will contribute storm water to the stream during the first hour of the storm. The density of stream channels in the catchment and the overall shape of the basin will therefore influence the routing of storm water and the shape of the flood hydrograph.

A high density of stream channels (*i.e.* km of channel per km² of basin) will produce a 'flashy' hydrograph with a short lag between storm and the response of the stream. An elongated basin will produce a flat topped hydrograph and so on. These points are illustrated schematically in Fig. 4.17.

The simplicity of this infiltration excess overland flow model, however, makes it of limited use. In semi-arid areas with limited vegetation cover and high rainfall intensities, or on highly impermeable catchments such as those underlain by permafrost or clays, it is of some value. Unfortunately, in the British Isles, for example, infiltration rates of soils commonly exceed all but the most infrequent and heavy rainfall intensities. Yet as we are all aware from news reports, streams flood quite often! We must therefore ask ourselves if there is another factor which prevents water entering the soil. One answer is if the ground is saturated already. If it is, saturated overland flow can occur.

What controls this saturated overland flow? Firstly the pattern of soil moisture at the start of the storm, *i.e.* the antecedent moisture conditions, how much rain has fallen over the catchment in the period before the storm. Secondly there is the pattern of soil moisture which develops during the storm.

Both of these are connected with the shape of the slopes and the pathways of moisture. Sites at the base of slopes and in hollows will become saturated sooner from rain falling on them *and* of course from water passing downslope through the soil. Such saturated sites are known as *contributing areas*, *i.e.* areas which contribute water to the flood hydrograph. They are obviously dynamic and changing, depending on the antecedent moisture conditions and the characteristics of the individual storm.

This is a simplified account of the *dynamic watershed model*. It recognises that the areas of overland flow change with varying soil moisture conditions and that this is not merely a function of the storm's intensity and the infiltration capacity of the soils. Some sites become saturated sooner and begin contributing water as storm runoff to the stream before others. Such areas change during the storm itself and vary seasonally too. As Fig. 4.17D shows these areas can be tracts of flat summit connected by pipe systems to the stream. In extreme cases the contributing area may be the stream itself, in other words rain falling on the channel alone produces the flood peak!

In the case of small basins therefore if we ask why the stream floods the answer that it's 'because it rains' is too simple. Our explanations need to take into account the actual workings of the catchment as a system. The geomorphological significance of floods and their human significance in terms of water management will be mentioned later in the chapter.

A THE FLOOD HYDROGRAPH

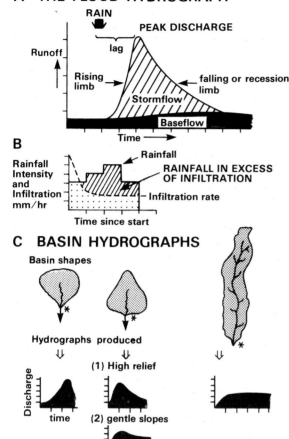

B

C BASIN HYDROGRAPHS

D DYNAMIC WATERSHED MODEL

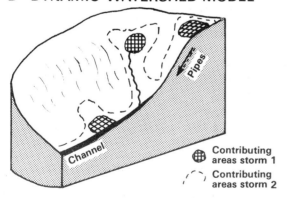

4.17 Floods

THE ACTION OF WATER ON HILLSLOPES

Before looking at the role of water in the channel in terms of its erosional, transportational and depositional work it is important to note that channels are only a small part of the basin area. Stream channels may be permanently occupied by water or in some situations they may be dry for part of the year or for a period of years. Whether permanent or ephemeral, however, it is useful to remember that they are only like the veins of a leaf, or to use a more accurate analogy they are gutters for the whole drainage basin surface. Soil is essentially rock material on its way to the sea! The processes of weathering and mass movement at work on slopes result in a supply of solid and dissolved material to the streams. These processes were discussed in Chapter 3, but running water itself can have effect on slopes, as we shall show.

Water erodes, and transports weathered debris from hillslopes in three main ways—by raindrop impact, by sheet flow and by the growth of rills and gullies. The *impact of raindrops* is rather like subjecting the hillslope to miniature bombs, the actual force of impact of the raindrop ejects debris after its impact. It has been found that 4 mm stones can be moved a distance of 20 cm and finer clay particles up to 150 cm. On level ground the particles dislodged are exchanged randomly, or shuffled around. On a slope this process results in a net slow downhill movement of particles. The actual mechanical impact of raindrops in some cases also seals the surface, making it more impermeable reducing infiltration and increasing overland flow.

With certain combinations of rainfall and surface characteristics *sheetflow* may occur. This is a thin film of water which moves down the slope, often as intermittent surges around vegetation. Its velocity and depth may be small yet the total flow may be large. Whenever such a moving film of water has sufficient force it can physically remove particles downslope. Sheet erosion is greatest on long steep slopes with low infiltration capacities. The force depends upon the water's depth and velocity, a function of infiltration and the fact that the greater distance downslope the more water there is to be moved. Steeper slopes increase the velocity and hence the potential erosion.

Both splash erosion and sheetflow are considerably hindered by the presence of a dense vegetation mat on the surface. The erosion of material from hillslopes will thus increase as precipitation increases, but after a certain point the precipitation results in a progressively denser vegetation cover which inhibits erosion. The removal of vegetation cover is therefore an extremely significant factor in explaining accelerated rates of erosion (soil erosion).

As water moves down the slope sheetflow gradually becomes organised into a series of *rills*, small channels a few centimetres deep, which join and separate. Some of these become deeper than others and last between storms. They transmit water from larger upslope areas. The larger the channel the less energy is lost by the water in overcoming friction. More energy is therefore available for picking up material, for deepening the rill into a *gully* and for extending it headwards. In an extreme form, *i.e.* on clays in semi-arid areas, gullying produces 'badlands' where the topography is finely and deeply dissected by gullies only a metre or so apart (Fig. 4.18). This type of rapid hillslope erosion can be seen on embankments and spoil heaps in humid areas before vegetation covers the slope.

These rills and gullies transmit material and water downslope towards the permanent stream network. It has been argued that the convex–concave slope profile reflects the operation of these processes. The upper convexity exists where soil creep and unconcentrated wash occur, the slope steepening so that the increasing amounts of material and water can continue to move downslope. The lower concavity reflects the operation of hydraulic processes. Discharge increases progressively downslope, leading to lower frictional losses so that the gradient can decrease yet still enable the

4.18 Badlands topography, a dense rill network dissecting soft sedimentary rocks, South Alberta

channel to transmit all the water and material being supplied to it. Rills and gullies are therefore in a somewhat intermediate position between splash and sheetflow on the one hand and the work of water flowing in permanent defined channels on the other.

Finally it is important to remember that water moves through the soil layers as well as over the surface. Water flowing in pipes, moving relatively rapidly, can carry weathered material downslope and the products of solutional attack can also be removed by throughflow, interflow and deeper seepages.

PROCESSES IN THE CHANNEL

'All soil is rock on its way to the sea' is a dramatic phrase summarising the hillslope processes mentioned in the last section and in Chapter 3. Slopes contributing both solids and dissolved material are an important part of the basin system. It is now appropriate, however, to look at the work of water in the stream channel itself.

Water flowing in the channel is an *agent of erosion, transport and deposition*. These three processes will initially be discussed separately but in streams the processes interrelate as we will see later. On a rocky stream bed the impact of flowing water alone, a kind of dragging action, is capable of smoothing the bed and removing material (*hydraulic action*). At waterfalls and rapids, with very high velocities only, *cavitation* can occur whose repeated hammer-like shocks can remove rock. Where rocks are susceptible to chemical attack the water can remove material in solution, a process known as *corrosion*.

Water unarmed with rock debris is relatively rare. Whirling eddies around obstacles in the bed or bank cause a kind of lifting action. Loose grains are sucked upwards and moved downstream. This *entrainment* of grains is caused by different water velocities and flow directions within the channel.

Figure 4.19 illustrates the relationships between particle size and velocity of water. At the top of the graph, line A is the *critical erosion velocity*, the lowest speed of water at which grains of a given size, loose on the bed of the channel, will move. Not surprisingly heavier cobbles and even larger boulders require very high velocities to move them. Pebbles move at intermediate velocities and sand-sized material can be quite easily entrained. Clays, however, require higher

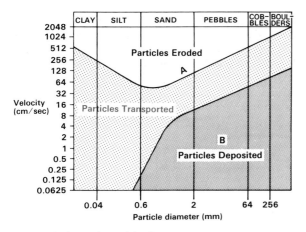

4.19 Velocity and particle size

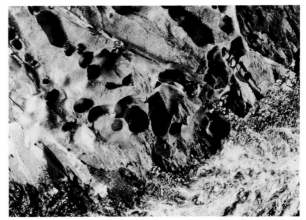

4.20 Potholes exposed in the bed of the Claerwen River

velocities to lift and erode them. Such fine particles form a cohesive and smooth bed resistant to erosion.

The actual velocity of water reflects discharge and channel slope and small discharges, in headwater tributaries, for example, are only capable of moving larger particles on steeper gradients. In the lower reaches of streams, although the gradients are usually less, the proportion of stream energy lost in friction with the bed is lower so the velocity will normally equal or exceed that in the smaller upland streams. This fact (which runs counter to impressions of rushing mountain torrents and slow majestic flows as the river enters the sea) is significant when the river's ability to transport its load of sediment is considered.

Figure 4.19 also shows in area B the *deposition* of particles. Between the critical erosion velocity curve and the deposition curve is the zone where particles are *transported* or carried. Particles require higher velocities to entrain them than to actually carry them but once velocity falls below a certain point the particles are dropped or deposited. This settling velocity is high for the larger and heavier particles, but for clays it is extremely low. A pebble with a surface area of $3 \, cm^2$ broken into clay-sized particles would have a surface area totalling $30,000 \, cm^2$ and the myriad particles would almost float in the water.

Once entrained the load is actually moved in two ways, either as bedload or suspended load. *Bedload* is defined as that part whose immersed weight is carried by intermittent contact with the bed. The particles move by combinations of sliding, rolling (especially if rounded) and a jumping motion known as salta-

tion. Bedload moves along the channel bottom slower than the actual velocity of the water. This material bumps and grinds into the bed and sides of the stream causing vertical and lateral erosion. By rolling and grinding against itself particles are rounded and fractured into progressively smaller sizes. Bedload pebbles may be spun round by water eddies and drill potholes in the bed of the stream (Fig. 4.20).

Suspension load is that part of the moving load whose weight is carried by the fluid. As Fig. 4.19 shows, once finer sands and silts are entrained they need little energy to carry them. Such particles besides being picked up from the bed may also originate from

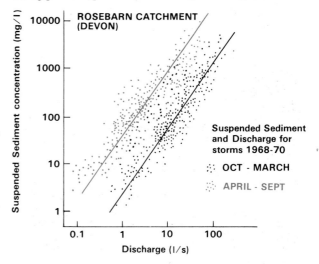

4.21 Suspended sediment and discharge

4.22   Lynmouth, the aftermath of the 1952 flood

the hill slopes above the stream channel. In most cases suspended load accounts for 70–90% of a river's total load of solids.

During *floods*, with increased discharge and velocity, the ability of a stream to carry material is dramatically enhanced (Fig. 4.21). The scatter of discharge and sediment figures for individual storms shows a positive relationship for both summer and winter seasons. (The reasons for the seasonal differences are examined in the exercises.)

Floods are also relevant when discussing bedload. Figure 4.22 is a photograph of the aftermath of the August 1952 flood at Lynmouth. During the storm 20,000,000,000 litres of water fell over this 100 km² Exmoor catchment and the discharge reached 511 m³/s, a figure only twice exceeded by the Thames this century in a basin 100 times larger. In total about 100,000 tons of boulders were moved into the lower valley. The kind of hydrological detective work necessary to reconstruct floods in such ungauged catchments solely from the evidence of their destruction is mentioned in Newson (1975).

Finally before leaving this general discussion of stream processes it is important not to become mesmerised by the visible signs of erosion and transport. The hidden giant in basin processes is the work of chemical processes and the removal of material in solution. Figure 3.29 indicated this for the Karkevagge area and mention has also been made of limestone solutional processes in Chapter 3.

The treatment so far has been in general terms. In the remainder of this section the focus will be on parts of the basin where the interaction between processes and forms can be seen. The first example is the delta environment where deposition is the theme.

Deltas: processes and forms

Deltas can be lonely evocative places, the seemingly huge sky reflecting level horizons, with a fragile and changing interplay of land and sea—truly the ends of the earth. On the other hand they may teem with people, like the deltas of the Ganges and Nile.

The term delta is applied to the low-altitude near-level plain of almost any shape formed by the river at its mouth. The Mississippi delta is the most well-documented, covering an area of 26,000 km². The river itself drains an area of 3,220,000 km² and each year it carries 590 km³ of water into the Gulf of Mexico laden with 450,000,000 tons of sediment! The section which follows is an attempt to show what happens to that sediment (40% of which is silt-sized and 50% clay-sized) as it nears and reaches the sea.

The delta has a complex history, having occupied seven delta lobes over the past 5,000 years. Figure 4.23 show these earlier lobes and the Balize delta, which is the focus of present deposition. The volume of material involved is immense, the whole delta containing 33,400 km³ of sediment of which only 110 km³ has

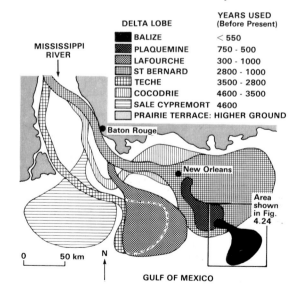

4.23   Recent delta lobes of the Mississippi

been deposited in the Balize lobe since AD 1500.

When the river reaches the sea its velocity is checked, the coarser debris and bedload is dropped to form the distal *bar*, marked by a shallowing from the 10-metre depth of the river channel to 3 metres or less at its mouth. The finer silts are carried further into the sea falling to form the massive submarine extension of the delta. The finest clays form a plume of sediment jetting out from the river mouth. Under freshwater conditions the clays' settling rates are extremely slow but salt water causes the particles to flocculate (join together) and drop to the sea bed.

At the sides of the channel *levees* are formed, especially during periods of high river discharge. The reduction in velocity at the edge of the channel leads to sedimentation and the build-up of these natural embankments paralleling the channel. These levees extend outwards (downstream). This can be clearly seen in Fig. 4.24 in the case of the Southwest and South Passes, and of course it is important to remember that they are growing at their tips beneath the sea. This is one method of delta expansion and growth.

Behind the levees sediment is deposited during the periods of overbank flows associated with high discharges. Reeds and swamps colonise the area and such vegetation slows the floodwaters and further encourages sedimentation, whilst the dead plant remains

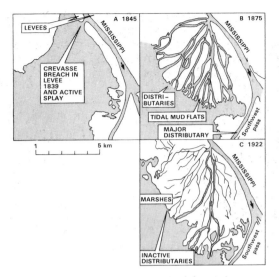

4.25　The evolution of a subdelta or crevasse, West Bay

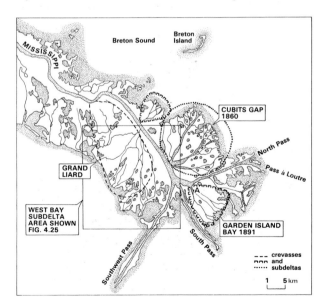

4.24　The modern delta of the Mississippi

4.26　The Mississippi delta, looking south above the Head of Passes. South Pass lies centre-right and Pass à Loutre swings off to the left. The levee breach at A was the site of the 1891 crevasse splay shown in Fig. 4.24

accumulate as peats.

In the river channel itself subdivision occurs. Sediment dropped in the bed forms shoals, the river diverts on either side of the shoal which builds by further accretion of sediment in the slower flows around it. It is colonised by vegetation and becomes a more durable element of the channel shape. The higher velocity filaments of water flow either side of it and levee extension now occurs on two advancing fronts. This can be seen in Fig. 4.24 in the case of Pass a Loutre on the east of the delta. In this way a growing network of distributaries develop as the channel splits and splits again.

This levee extension is however only one mechanism of delta growth. *Crevasses* or subdeltas form most of the dry land present in the area. A typical crevasse is the West Bay Complex shown in Fig. 4.25. At the time of the first surveys a shallow bay existed in this area to the west of the Mississippi itself. In 1839 a break occurred in the levee and by 1875 (Fig. 4.25B) we notice a system of distributaries splaying out 16 km from the crevasse breach, a great fan of debris dropped by successive floodwaters. The 1922 map shows little further change apart from a southerly extension of some of the channels and their associated deposition. Many of the others had become virtually abandoned. They had been so successful in building seawards that this crevasse break had no gradient advantage over the Mississippi River Pass itself, the underlying rationale for the crevasse subdelta in the first place. The main levee of the river becomes rebuilt and other crevasses open to fulfil this relief function (see Fig. 4.24).

Now deprived of sediment the West Bay crevasse begins to change. It subsides, the weight of material splayed out during the previous active phase depresses the unconsolidated sediments of the earlier lobes. Lakes expand and in time become rounded by wave action (see Fig. 4.25C). Meanwhile the outer edge of soft and unconsolidated sediments are reworked by the sea forming continuous barrier beaches. The rate of coastline retreat is quite rapid, in some exposed parts of the delta it reaches 20 m a year. If you look at the north-east corner of the map you will see Breton Island, which correponds to the line of the earlier St. Bernard lobe (shown in Fig. 4.23).

We therefore have a *cycle of evolutionary development* of the crevasse subdelta. New channels are born, land is created, channels abandoned and sediment-starved areas sink below the waves. In the case of the

VARIABLE	DELTA GROWTH	
	SLOW	RAPID
SEDIMENT INPUT	LOW	HIGH
WAVE ENERGY	HIGH	LOW
TIDE RANGE	HIGH	LOW
SUBMARINE GRADIENT	HIGH	LOW
SPEED OF VEGETATION GROWTH	LOW	HIGH
SUBSIDENCE	HIGH	LOW

4.27   Variables in delta growth

older lobes, subsidence and marine reworking are not the only processes leading to loss of land—the post-Pleistocene sea-level rise has also been a factor. The delta is thus a dynamic and changing environment but as the pivot of America's inland waterway system the river isn't allowed to go its own way (possibly flowing to the Gulf via the Atchafalaya River). Its banks are strengthened and bed dredged so that the ports of New Orleans and Baton Rouge can continue to be used.

It is a common misconception that deltas only grow in tideless seas. True the Mississippi delta has a tidal range of 0·45 m but the Ganges–Brahmaputra, twice the area of the Mississippi, has a normal tidal range of 5·7 m. What really matters is the relative balance between the input of debris on the one hand and the subsidence beneath the sea and the reworking of sediments by the waves on the other. Figure 4.27 summarises these points. Finally, although this discussion has been in terms of the Mississippi and although deltas come in a range of shapes the processes reviewed here are applicable elsewhere. The depositional processes are also relevant in the examination of upstream areas which forms the focus of the next three sections.

Meanders

Meanders, the curving bends in a river channel are one of the most widespread of fluvial landforms. Occurring in streams of all sizes and at all elevations they reflect elements of fluid flow and the processes of erosion, transport and deposition within the channel.

Figure 4.28 is a model of an established meandering reach of a stream. The items at the top of the diagram are the parameters of meander dimension. Towards the bottom information on the channel depth, cross section and water movement is shown. As with any channel, water at the surface is swifter than at the bed and this maximum velocity vector (the blue arrow) moves towards the concave bank in the bends. Since water cannot 'pile up' with a sloping surface, water at the concave bend and slightly downstream from it has a downward motion as well as a downstream one (cross section B). The channel cross section is asymmetrical here with erosion on the concave bank and alongside the deeper pool. Deposition occurs in the slower moving water on the inside of the bend. Moving downstream the maximum velocity vector swings across the channel, the inflexion being marked by a shallowing bed, and then impinges on the concave bank of the next bend. Although the maximum velocity vector swings in this way it is important to note that there is a circulatory movement across the channel (section inserts A and B). Any molecule of water would therefore spiral in three dimensions, a screw-like or helical flow, as it moved downstream.

Meanders have been a recurrent theme of enquiry in geomorphology. As a result of this a number of relationships have been discovered and a selection of these are shown in Fig. 4.29. Meander length, for example, has been found to be about ten times channel width. This is shown in Fig. 4.29A—stream X, for example, has a channel width of 10 and a meander length of 98. As the graph shows, this relationship holds with streams covering a range of sizes (shown by the linear band made up of the other 47 streams plotted). Meander width has also been found to be 14–20 times the channel width although this is not shown here. In other words the form itself (the meander) is repeated irrespective of the actual size of streams.

Figure 4.29B shows the relationship between meander wavelength and discharge, a process-form relationship. Figure 4.29C shows the relationship between discharge and channel slope. The streams have been differentiated according to their habit, whether they meander (blue) or braid. The latter channel form will be discussed later but the graph shows that for streams of any given size, or discharge, meandering occurs on lower bed slopes than braiding.

Finally Fig. 4.29D is a plot of channel patterns and the locally available bed materials. Stream Y, for example, is flowing over material which is 40% sand-sized, 50% clay-sized and 10% solid rock, and it has a meandering channel. Stream Z flows in 95% sand and has a braided channel. The zone in which meandering occurs is indicated in blue. Meandering doesn't develop in hard resistant bedrock where channels are straight or crooked, or in sands which are

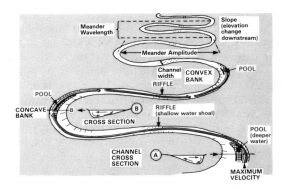

4.28   Meander shape and water flow

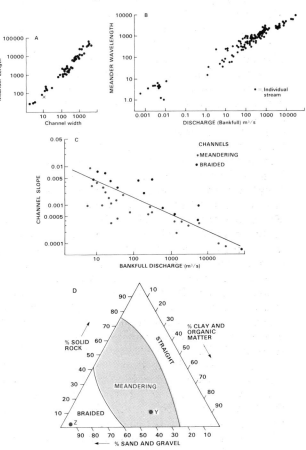

4.29   Meanders
A   Meander length and channel width relationship
B   Meander wavelength and stream discharge
C   Discharge, slope and channel habit
D   Channel patterns and bed material

so mobile that the shape of the channel will change with every variation in discharge. Meandering channels are therefore best developed in mobile materials which can be eroded, yet which have sufficient cohesion to maintain their shapes in periods of low flow.

From these sort of observations it is concluded that meanders, their size and shape, are governed by the interaction of water and sediment in the channel. They are thus in a kind of quasi-equilibrium with discharge, slope and bed materials.

A number of *theories of meander origin* have been put forward, some based on laboratory work with stream tables, some on mathematical and theoretical models and some on field deduction. Riffles and pools, alternating shallows and deeps in the bed (shown in Fig. 4.28), are characteristic of alluvial streams, whether meandering or straight. Where the stream is straight, turbulence occurs over the riffles, resulting in high energy losses, and smoother flow over the pools. This means there is an unequal distribution of energy loss in the stream. In the meandering situation the curve in the channel over the pool increases resistance and thus energy loss. Meanders therefore equalise energy loss, produce equilibrium if you like, within a section of stream. Like any open system the river behaves in such a way as to minimise the variations in its properties by distributing them (*i.e.* energy loss) throughout. Secondly the actual shape of the curves themselves (sine generated curves), by keeping the rate of change of curvature to a minimum (minimum variance) reduces the overall energy loss to a minimum. Thus it can be proved theoretically that once the stream is not straight the most *probable* shape is the meander.

A second group of theories relates to the way in which water flows. Earlier mention was made of helical flow in the established meander. This appears to be an inherent feature of fluid flow, originating with turbulence and resistance. In turn it influences the distribution of velocity in the stream and hence where erosion and deposition will occur.

The whole topic is complex and no single theory is satisfactory, certainly not chance obstructions like fallen trees! In laboratory stream tables a straight and uniform channel develops a series of pools about five times the channel width apart. These then migrate sideways, forming alternate pools in different directions, the channel lengthening and curving to give an average wavelength of two interpool spacings, or 10 bedwidths, as in nature. In other words the irregularity appears in the long profile of the channel first (pools and riffles) before there is any plan expression in the form of meanders.

Flood plains

Flood plains are flat strips of land bordering river channels, composed of unconsolidated deposits, which are periodically inundated. To take up this latter hydrological part of the definition focuses attention onto a basic question, 'why isn't the channel big enough to take the extra water?' Figure 4.9 showed that light rain occurred more often than moderate falls and the heaviest rainfall is even more infrequent. On most days therefore the river has only a moderate amount of water in it. On a few days of the year it rises to just fill the channel (*bankfull stage*) and more infrequently it overtops its banks. Most river channels are shaped by and adjusted to moderate flood flows and not the larger infrequent discharges. Figure 4.29B is an example in the case of meander wavelength and bankfull discharge.

In the discussion of meanders brief mention was made of erosion and deposition in the meandering reach. Figure 4.30 shows changes in a part of the Mississippi's course over a period of 83 years. The meander has *migrated* downstream with a combination of erosion on the concave bank and deposition on the convex. Such a migration 'planes off' any higher land adjacent to the river, leaving behind a low cliff, or *bluff line*, bordering the meander belt.

During this migration *point bar accretion* occurs on the convex bank. Water is slower moving here and of course the helical flow component is upwards with water decelerating. Reduced velocity results in deposi-

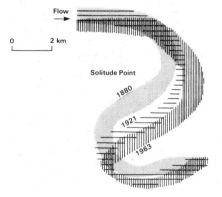

4.30   Meander migration on the Mississippi

tion, initially, of the coarser material of the bedload at a point downstream from the axis of curvature. As the bar grows, deposition of finer sediment occurs in the shallower and calmer water between it and the convex bank. Vegetation may also encroach, further trapping sediment and building up deposits of dead organic matter. As the meander shifts it leaves a micro-relief of swales (hollows of finer sediment) and scars (previous bars). This process is shown in Fig. 4.31.

Besides migrating downvalley meanders may also migrate laterally and the meander belt may at any one

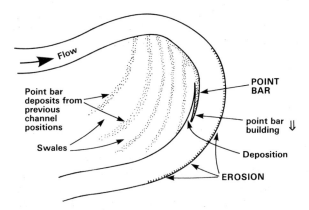

4.31   A point bar

time only occupy a part of the flood plain. When this situation exists *levee* development may be a little more obvious. The formation of these wedge-shaped low ridges of sediment bordering the channel has already been introduced in the discussion of delta development. Levee breaching can also occur on the flood plain and a crevasse splay or very low angle fan of material may form, progressively thinning and becoming finer as distance from the channel increases. The river may also overtop the levee in times of flood. Bedload is deposited close to the channel and the suspended load settles in the slow-moving waters over the flood plain to form the overbank deposits. It is still a common misconception that flood plains are formed entirely of such overbank deposits. However, point bar accretion, meander shifting and the general reworking of sediments in the channel are more significant processes as they take place continuously. The Mississippi, for example, reworks about a third of its flood plain every fifty years (see Fig. 4.30). The flood plain can therefore be viewed as a store for sediment on its way to the sea. Between 80 and 90% of this material has been placed in storage by lateral accretion and only 10–20% by the overbank veneering during floods.

At periods of high discharge where two concave (*i.e.* eroding) banks are adjacent the river may break

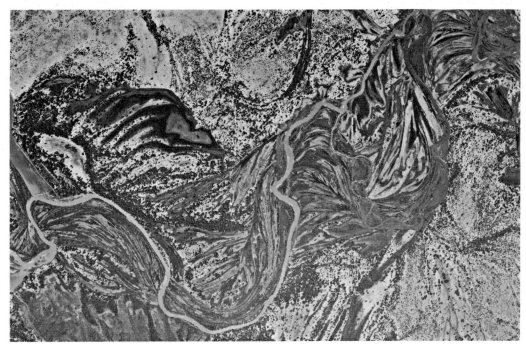

4.32   A vertical aerial photograph of a flood plain in West Africa. The migrating channel of the river has left a complex legacy of cutoffs, bars and swales

through the meander neck forming a *cutoff*. For a while both channels may be used but the entrance to the loop soon becomes plugged with sediment leaving the former course abandoned as an oxbow lake. Like all lakes it is only temporary, filling with flood sediments and the growth and decay of swamp vegetation, but it leaves behind a scar easily detected by vegetation differences on air photographs. In some situations a chute cutoff occurs when the river breaks through its bank to follow the line of a swale in the point bar deposits.

Figure 4.33 is an example of this cutoff process. It is reasonable to ask if meanders can migrate freely (as in Fig. 4.31). Why should a meander 'catch up' with its downstream neighbour? The migration rate is, in fact, affected by bank resistance and where a meander encounters more resistant bed and bank its migration slows. Figure 4.33 shows the existence of two clay plugs, remnants of an earlier cutoff, which can be envisaged as retarding the meander so that the upstream meander loops could begin to impinge on it, thus forming a neck and cutoff. Finally mention should be made of another feature of the flood plain. Streams draining the flood plain may find entry to the main stream difficult because of the levees and they may then flow down the valley as deferred tributaries.

Flood plains are thus a product of and a functioning part of the whole stream environment. They play a part in the adjustments that a stream makes to variable quantities of water and sediment which it stores and carries to the sea.

Braiding

Mention has already been made of braiding channels

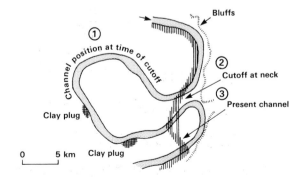

4.33   Mississippi River: False River cutoff

4.34   A braided channel, the South Saskatchewan River in the Rockies

in connection with Fig. 4.29. Defining a braided channel is not easy and Fig. 4.34 illustrates its main features. The overall course is relatively straight, but there is a wide shallow bed choked with bars and there are frequently rapid shifts in the positions of the bars and channels. At periods of high discharge the Yellow River in China shifts laterally up to 120 m a day, for example. Braiding tends to develop in material which is coarser than meanders (the triangular plot Fig. 4.29D indicated this), with streams which have high bedloads, higher channel slopes and irregular discharges. Such conditions are found in glacial meltwater streams and in semi-arid environments.

THE 'AVERAGE' RIVER: A SUMMARY REVIEW

Figure 4.35 shows the relationships of width, depth, velocity, suspended load, channel roughness, river slope and bed material size to discharge. The relationships are shown for two stations, one upstream (A) and one at the mouth (B). The relationships in time of flood are shown in blue, at times of low flow in black. The diagram summarises the complete *hydraulic geometry* of a channel system and gives an opportunity to review some relationships which have been introduced in the sections above.

At the top of the diagram is the width-discharge graph. At low flow (black) the channel width increases downstream. This line has an upward slope to the right of 0·5, indicating width increasing at half the rate of the downstream increase in discharge. This relationship holds in flood conditions too, as shown by the blue line. The increase in width at the stations from low flow to flood is shown by the broken black line.

Suspended sediment load also slopes up to the right (0·8) showing the increase in sediment is slightly less rapid than the downstream increase in discharge. Mention has been made earlier of the tremendous increase in the stream's abilities to carry material in times of flood. This is indicated by the displacement of the blue line representing flood conditions.

Velocity is shown to remain almost constant (or show a very modest increase) in the downstream direction. Stream velocity naturally increases in times of floods but this downstream relationship still holds. At the same time slope (energy) decreases from source to mouth, reflecting the upwardly convex curve of the river's long profile. These two facts—velocity constant but slope decreasing—can be reconciled if we look at channel width and depth. Both of these increase downstream, so that the channel becomes more efficient. Channel roughness also shows a gentle drop downstream and there is also a decrease in bedload size (lowest graph). These will lead to a downstream reduction in flow resistance. Bars and meanders in the lower course, however, offer some resistance to flow and partially offset the advantages of a decrease in resistance associated with the smaller particle size of the bedload.

Figure 4.35 is of some assistance in helping us understand how the river adjusts to the increasing discharge in the downstream direction as progressively larger areas of land are drained. Simplistically the river could merely increase its width downstream, or its depth or, of course, its velocity. Alternatively it could combine the various options in a host of ways. In fact as studies from many rivers have shown (and these are summarised in the diagram), in spite of the range of environments in which streams flow and the range of sizes which they reach, there is a considerable degree of consistency in the way in which the river copes with the increase in discharge by adjusting its channel form.

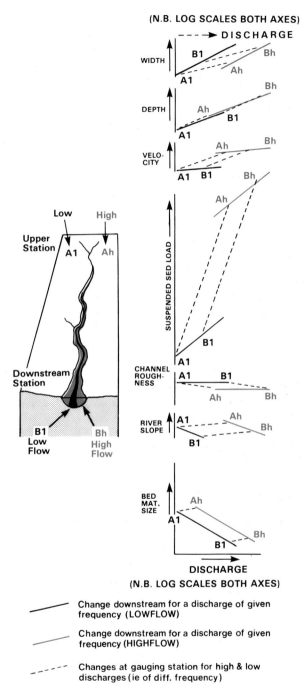

Change downstream for a discharge of given frequency (LOWFLOW)

Change downstream for a discharge of given frequency (HIGHFLOW)

Changes at gauging station for high & low discharges (ie of diff. frequency)

4.35 River channels: Average hydraulic geometry changes downstream and in flood (see text for explanation)

THE DRAINAGE BASIN NETWORK: PATTERN, PROCESS AND TIME

Until now the drainage basin has been looked at very much in terms of its elements, the channel and the processes within it. The drainage basin will now be considered on a somewhat larger scale, both aerially and in terms of its evolution.

The ubiquitous veination of river channels has long attracted the attention of earth scientists who have attempted to analyse and explain drainage patterns. Figure 4.36 is an example of the drainage pattern of

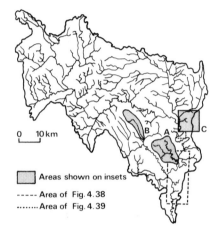

4.36   The Wye basin, drainage patterns

the River Wye. Such patterns have been described in terms of *analogs*, *i.e.* forms which they are similar to. A tree-like pattern of a master stream and its branching tributaries has been called a *dendritic* pattern. An elongated basin with a master stream and many single tributaries resembling a feather has been termed a *pinnate* pattern. A *trellis* pattern is formed where relatively straight tributaries join in a right-angled grid-like pattern. Streams draining away from a single central point form a *radiating* pattern and where a circular arrangement of streams can be seen it is described as an *annular* drainage pattern.

When applying such an approach a considerable degree of *structural control* of the stream pattern is assumed, either overtly or by unstated assumption. On a surface composed of homogeneous rocks, *i.e.* with no differences of resistance, a dendritic pattern may develop. Figure 4.36A is an example of this where the

Garren Brook has a tree-like pattern flowing on Devonian sandstones and marls. Where strata is gently dipping a parallel series of major streams may have developed and the pattern within the sub-basins may be pinnate (Fig. 4.36B).

Where rocks of uneven resistance existed close together streams were envisaged as exploiting softer rocks or lines of weakness. The latter may be fault lines in massive crystalline rocks and Fig. 4.37 illustrates

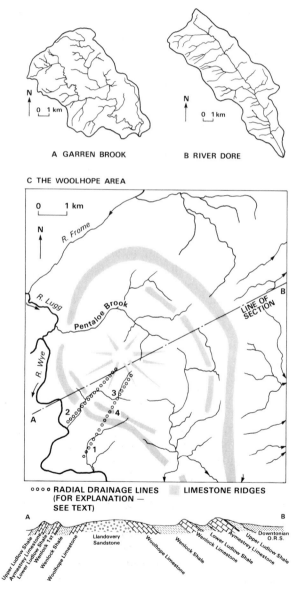

this. An example of the former is shown in Fig. 4.36C, the drainage pattern in the Woolhope area of Herefordshire. Here we have a pear-shaped dome of Silurian sedimentaries (alternate layers of shales and limestones) 'punched through' the surrounding younger Devonian Old Red Sandstone series (marls and sandstones). The original dome may have initiated a radial pattern of streams. The term applied to such streams flowing in response to the initial surface is *consequent*, *i.e.* their direction was consequent upon the form of the initial surface. Parts of the present-day pattern also have elements of a radial orientation, and this is shown in the diagram.

The dome would have been subjected to weathering and erosion, the streams cutting valleys into the structure. Thus they would have gradually exposed the older rocks beneath, the alternation of shales and limestones with the Llandovery sandstone in the centre of the dome. The shales are relatively soft and have been exploited by tributaries. These tributaries exploiting lines of weakness are known as *subsequent* streams. They flow along the strike of the outcrops. In situations where the outcrops are not as tightly curved as in Fig. 4.36C such subsequent stream development will give rise to the trellis pattern mentioned earlier. In the case of the Woolhope area the subsequent streams deepening and widening the shale vales have formed elements of an annular pattern as the drainage adjusts to the structure and the lithological variations. The inner-facing limestone scarps are shown in Fig. 4.36 together with a generalised geological section.

Before leaving this small-scale example mention should be made of another process which occurs as rivers adjust their patterns to structure. A subsequent stream exploiting the less resistant rock may expand headwards so that it *captures* the adjacent headwaters of one of the original consequents. Some of the original radial streams would have been more powerful than others. This would have enabled them to deepen their valleys, leading to further lowering of their tributary subsequent streams. Evidence for this in the Woolhope area is conjectural but sufficient to illustrate the process with drainage lines 1 and 2. In this case line 2 was more aggressive and its tributaries expanded eastwards to capture the headwaters of drainage line 1. At 3 we have the *'elbow of capture'* where the line of the original consequent stream 1 becomes incorporated into the expanded catchment of

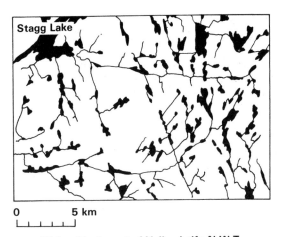

0   5 km

70 km Northwest of Yellowknife N.W.T.

4.37 Rectilinear fault-guided drainage on the Canadian Shield

stream 2. A small notch in the summit of the Wenlock limestone scarp at 4 may represent the line of consequent 1. Such gaps are known as *wind gaps*. Where their orientation and gradient relates to the original line of drainage and when they are littered with stream deposits the previous existence of the stream can be reasonably assumed. Downstream in the beheaded consequent the stream may appear to be rather small for its valley and it is known as a *misfit*. It is emphasised that this account for the Woolhope area is conjectural. In the further readings examples will be found of river capture where stronger evidence exists.

Examples, however, may be found where such adjustment to structure has not occurred. Streams initiated on an earlier surface, a depositional veneer or an erosion surface, have etched themselves vertically and laterally into the underlying rocks as relative sea–land levels have changed. During this process they may completely ignore geological differences. In the case of the River Wye, between Kerne Bridge and Monmouth it crosses and recrosses the Devonian-Carboniferous boundary four times and travels across sixteen lithological changes without any evidence of adjustment! In this area the River Wye has lowered its valley since the Pliocene. This has resulted in a series of meanders which have been cut into the more resistant block of the Forest of Dean plateau (Carboniferous rocks). Upstream, where the river flows over the Old Red Sandstone series, valley widening processes have kept pace with downcutting, open-

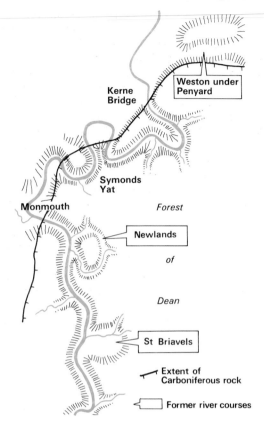

4.38   The lower Wye

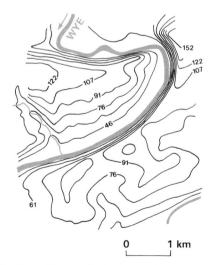

4.39   The River Wye, an incised meander at Ballingham

ing up the Hereford Plain in the central part of the Wye basin. The Wye, therefore, somewhat atypically leaves the uplands of Mid-Wales to meander across the Hereford Plain before the valley sides close in again for the last 60 km of its course.

In the lower section of the Wye are a series of *entrenched meanders*, indicated in Fig. 4.38. As the river has cut down it has also continued to migrate downstream. Ballingham spur (Fig. 4.39) is an example of such an *incised meander*. A process analogous to the meander cutoffs mentioned in the earlier section would also seem to have occurred and three *abandoned meander* courses of the lower Wye are shown in Fig. 4.38. The elevation of the oldest cutoff, the Newlands, is the highest and has been later modified by tributary streams downcutting to the lowering main river.

Where a drainage pattern is not adjusted to structure and if adjustment is assumed to be normal (a big if?) two broad explanations have been offered. The first is that of superimposition, *i.e.* when the drainage pattern is developed on an overlying surface of rocks which rest unconformably on a different structure beneath. The pattern is then *superimposed* on the underlying rock. In the case of *antecedence* the structure, such as a fold or fault, is assumed to be younger than the river. The rate of (or extent of) change was so small that the river was able to maintain its course during earth movement. The Brahmaputra transversely crossing the fold axes of the eastern Himalayas is an oft-cited example.

In view of the problems suggested above with the analog and evolutionary approach to drainage pattern analysis and description the objective *numerical description* of drainage patterns will now be introduced.

Quantitative analysis of drainage basins

Some of the techniques used in the numerical description of drainage basins are shown in Fig. 4.40. In terms of the actual network this involves initially an analysis of the basin's drainage in terms of stream *ordering*. Although the pioneer work was done by Horton the Strahler system is shown as it has been the most widely applied. In this system the unbranched tributaries are designated as first-order streams. Where two first-order streams join, the segment downstream becomes second-order and so on until the order of the trunk stream exiting from the basin is arrived at (Fig. 4.40).

A selection of other measurements which can be made of the stream network and the basin are also shown in Fig. 4.40. From what has been said earlier about the dynamic watershed model and contributing areas it is obviously important to specify on what basis the drainage network has been portrayed, *i.e.* if from a topographic map at what scale, from air photos or from field investigation.

In connection with stream ordering Horton discovered the *Laws of Hydraulic Geometry* which revealed an underlying and repeated organisation in the geometrical aspects of the drainage network. These are shown in Fig. 4.41. Usually a single basin is plotted on these semi-log graphs but the example shown here is data for seven basins on Dartmoor and 71 basins in the Unaka Mountains of Tennessee and North Carolina. Figure 4.41A shows the relationship between the number of segments of each order. The graph is a mathematical model, a negative exponential function, in which the number of stream segments of successively lower orders tend to form a geometric series, increasing according to a constant bifurcation ratio. This is the law of stream number. Stream length is plotted in Fig. 4.41B. In this case the mean length of stream segments of successive orders form a geometric series, beginning with the mean length of the first-order segments and increasing according to a constant length ratio. Finally area is shown in Fig. 4.41C where the mean basin areas of successive stream

orders tend to form a geometric series, starting with the mean area of all the first-order basins, and increasing according to a constant area ratio.

When basins diverge from these relationships (*i.e.* in terms of variations in the gradient of the lines representing differences between regions or in terms of lack of 'fit' of one element, such as the number of first-order streams) we might ask why. An explanation could involve the history of the basin, the geology of the catchment or elements of climate. The difference between Dartmoor and the Unaka Mountains is a case in point. Chorley and Morgan studied these two areas, both developed on granite. Their drainage density differs (3·45 for Dartmoor and 11·8 for Unaka) which accounts for the vertical separation of the plots. The explanation put forward in this case was a difference in rainfall intensity acting on these two areas since the Tertiary.

What underpins the regularity of number, length and area detected by such Horton analyses? Two explanations have been adopted, the rational and random. The *rational* model, developed by Horton him-

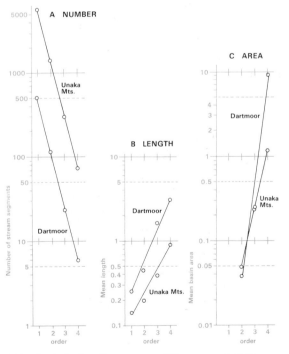

4.41 Basin hydraulic geometry ('Horton analysis')

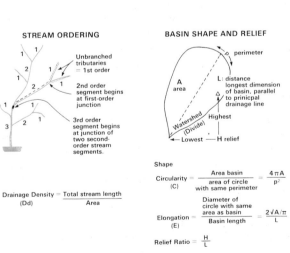

4.40 Basin measurements

self, is shown and explained in Fig. 4.42. The drainage
network is viewed as evolving from a series of parallel
rills on the hillslope. The network becomes organised
by successive stages, reflecting the operation of micro-
piracy (when one rill captures runoff of an adjacent
rill) and cross grading (when the movement of water
towards the rill produces a slope in that direction). The
diagram also shows that the units which are added
possess the same properties as the complete network
and form at a rate proportional to the size of the system as a whole (allometric growth). A model network
developed by these principals has the same geometrical qualities as those found in the natural landscape.

The alternative approach is shown in Fig. 4.43.
Drainage networks have been simulated by computers
which when programmed with certain constraints (*i.e.*
rivers can't drain upslope or flow in circles) can produce networks based on *random* principles. Such
networks represent the most probable state. Basins
simulated in this way, by random channel growth
and integration, have the same geometric properties
as those revealed by analyses of actual networks. To
demonstrate this point the figures for the basin in
Fig. 4.43 are included and could be inserted on Fig.
4.41A. In simple terms drainage networks in nature
therefore represent the *most probable* condition.

Drainage density (Dd) is perhaps the single most significant basin measurement which can be made. Some
caution needs to be expressed, however, in view of
problems of definition and measurement. What is a
stream? Is it a high-frequency overland flow line or
does it demand some specific topographic expression
like a channel? Different cartographers will have different conventions on how to represent ephemeral
streams and the scale of the map will affect the number

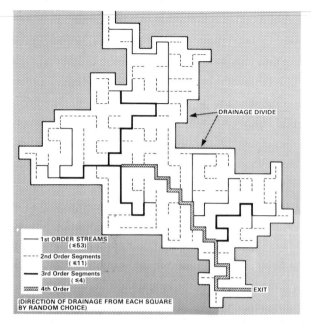

4.43   A simulated drainage basin network

of tributaries which are shown and the amount of
detail on sinuous channel courses. Care and consistency are obviously important guidelines.

Drainage density is an indicator of the distance
separating streams. This will affect the speed with
which the stream responds to storms—the higher the
density the faster the response—and the amount of
sediment which enters the stream from the hillslopes.
Figure 4.44 shows the Dd and mean annual flood relationship. In both areas plotted there is a positive
correlation but the regional variation is noticeable.

The input to the basin—precipitation—can be
assumed to be important in determining drainage
density. But precipitation isn't the same as runoff. A
more detailed approach is therefore needed which
somehow reflects the effective precipitation: that portion which is available for runoff after evapotranspiration losses. Figure 4.45 is an example of this approach
applied to part of the western USA.

In summarising a range of drainage-density studies
Gregory and Walling conclude that 'drainage density
reflects precipitation intensity', a relationship suggested earlier in connection with the data plotted in
Fig. 4.41 for the two granite areas of Dartmoor and

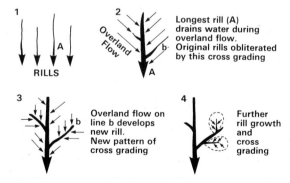

4.42   Development of drainage by rills and cross grading

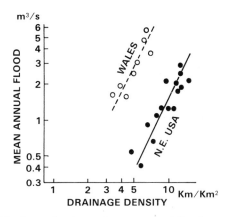

4.44 Drainage density and mean annual flood

the Unaka Mountains. High-precipitation intensity is characteristic of semi-arid areas which on the whole have the highest drainage densities, whilst the lowest occur in humid temperate areas. Local variations reflect such basin characteristics as ground cover, soil and bedrock. You might think of how these could be measured and related to drainage density—an interesting problem in research design!

There are only a small number of studies available of drainage network changes over *time*. One of these concerns the development of drainage networks on glacial till sheets in the American Mid-West. In this location are five plains of depositional material each successively exposed to fluvial processes as the ice sheet retreated. An analysis of the drainage of the five till sheets is listed in Fig. 4.46.

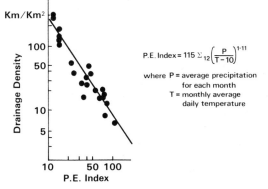

$$\text{P.E. Index} = 115 \sum_{12} \left(\frac{P}{T-10} \right)^{1.11}$$

where P = average precipitation for each month
T = monthly average daily temperature

4.45 Drainage density and P.E. Index for 23 basins in Colorado, Utah, Arizona and New Mexico

AGE	>40 000	40 000	17000	15000	13000
Dd	89	78	77	31	25
Σ 1st order	40	34	22	6	2
Σ 2nd order	14	11	6	2	1
Σ 3rd order	4	3	2	1	0
Σ 4th order	1	1	0	0	0

4.46 Drainage characteristics on five Iowa tills

Although the till sheets are naturally not of exactly similar lithologies, they do indicate how the drainage evolves. This is shown dramatically in the map (Fig. 4.47). The image of drainage density increasing with the length of time the till has been exposed to fluvial processes is confirmed in Fig. 4.48. The rate of change of drainage density appears most rapid in the first 20,000 years. It has been suggested that the levelling off in the rate of increase reflects the establishment of a kind of equilibrium as the basin areas become filled by the growth of streams. The relationship of channel length number over time (Fig. 4.48) follows the geometric laws used above to describe the relation of number to length when different regions were examined. Leopold, Wolman and Miller claim that this supports the view that the spatial distribution is in fact a 'growth' model.

Studies of changes over much shorter periods of time, during floods, are also somewhat rare. One of the few, undertaken in the context of a fascinating study of the floods on the Mendip Hills in 1968, was that of the Swildon's catchment by Hanwell and Newson. Figure 4.49 shows the drainage network and contributing areas under three conditions. The graphs of number/order and length/order show relationships

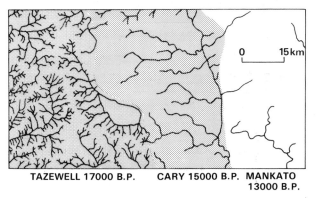

TAZEWELL 17000 B.P.   CARY 15000 B.P.   MANKATO 13000 B.P.

4.47 Drainage patterns on three Iowa tills

under three discharge conditions. They also show the situation on 10th July 1968, when discharge into the cave reached 3·47 m³/s and drainage density over the catchment was an almost incredible 47 km/km².

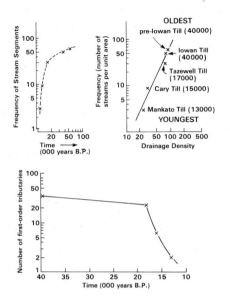

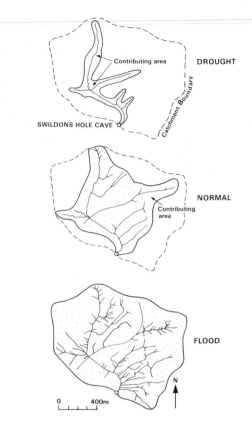

4.48   Drainage changes over time on the Iowa tills

The basin's three-dimensional form

The approaches outlined above do not consider the basin as an area with slopes. One way to do this is by constructing area height, or hypsometric, curves. These would indicate the existence of extensive summit areas or intermediate benches within the basin. A development of this is the *percentage hypso-metric curve* which enables basins of different sizes and height ranges to be compared. Their construction is shown in Fig. 4.50A.

Such hypsometric curves for basins reflect the interaction of slope-wasting and channel-deepening processes upon the underlying rock. Strahler suggested on the basis of inspecting many curves that the development of basins consisted of two phases. In early development, when the drainage system is still extending and not all the surface has become permanently integrated into slopes leading to the channels, curves with integrals exceeding 60% are found (Fig. 4.50B). This phase he termed 'inequilibrium'. With the expansion of the drainage network

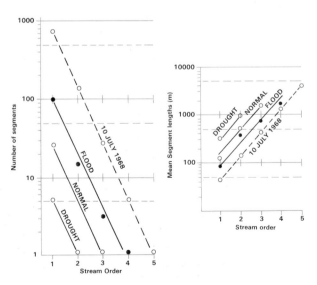

4.49   Drainage networks and discharge

and its contributing slopes the phase of 'equilibrium' is entered. Curves fall across the centre of the plot with integrals of 43%, or within the range 35–60%. As the slope and channel processes continue to remove rock materials Strahler suggested that the curve retained this shape although the total relief within the basin was reduced. This suggests that some kind of *steady state* is achieved. In other words hypsometric analysis illustrates that a *time independent* state is achieved in terms of the relative proportions of the basin. Although we are talking here of very long periods of time it does illustrate a balance between the endogenic elements (the rock and all its characteristics) and the exogenic elements (weathering, transport and erosion).

Strahler suggested the only exception to the equilibrium curve occurred when isolated residual hills remained within the basin (Fig. 4.50D) Hypsometric integrals would drop below 35%. Once the residual is finally weathered away the curve returns to the equilibrium position.

The Davisian model: the cycle of erosion

'Models are beautiful and man may be proud to be seen in their company. But they may have hidden vices. The question is after all not only whether they are good to look at but whether we can live with them' (Kaplan, *The Conduct of Enquiry*). Physical geography during the twentieth century has certainly lived with Davis's model of the cycle of erosion. In so doing it almost isolated itself from the mainstreams of earth science for almost half a century. The cycle of erosion is a *model* in that it identifies some supposedly significant aspects of the real world and associates these into a working system. Davis himself wrote 'the scheme of the cycle is not meant to include any actual examples at all'. The problem lay therefore not with the model, but in its uncritical application.

Davis's model of the cycle of erosion systematised earlier views and was a product of post-Darwin nineteenth-century ideas. The variety of natural land-

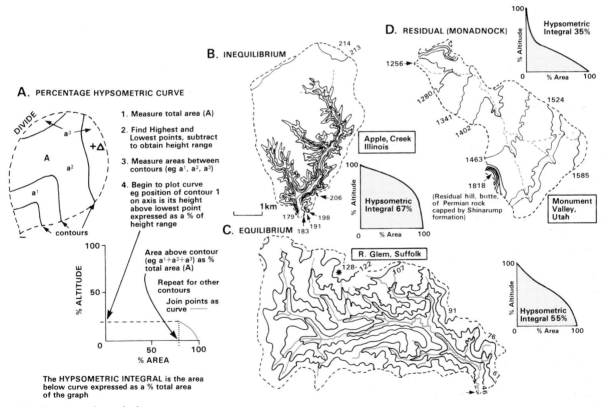

4.50 Hypsometric analysis

scapes were viewed as being related and forming part of an *evolutionary* sequence, from youth through maturity to old age. Although landscapes were analysed in terms of three variables, structure, process and stage (time), change through time was central to the argument.

The model assumed initial uplift after which drainage developed in a downslope direction on the flat surface. Streams rapidly etched themselves into this surface to produce V-shaped valleys with interlocking spurs. This was the stage of youth, marked by increasing relief as valleys deepened and widened to consume the original surface. Once this surface was totally consumed the landscape was at the stage of *maturity*. Streams lowered themselves more slowly, divides were consumed and slopes declined in angles so that relief became less. Streams became 'graded' to the base level of the sea, marked by smooth upwardly curved long profiles, and the development of meanders and flood plains was associated with lessening stream slope and supposedly reducing velocity.

After a long period the landscape was reduced to peneplain, a surface of low relief close to base level. The stage of *old age* was marked by extensive flood plains but occasional residual hills (monadnocks) may have broken the monotonous relief.

How valid is the cyclic model in explaining landscape? Davis's terms of youth, maturity and old age together with the snapshot images these convey have been long-lived shorthand amongst geographers. His ideas of rejuvenation, when sea-level lowered and rivers began vertical downcutting again, commencing at their mouths, has been a useful organising idea. Developing from this the idea of *polycyclic* landscapes, those arrested in their progress towards the peneplain by sea-level changes, has also been valuable. An example of this approach was hinted at in the earlier discussion of the lower Wye.

There are, however, a number of *assumptions*, such as rapid initial uplift of a land surface followed by a long period of stable base level (sea-level) which may be untenable. Secondly, are its assumptions about processes accurate? Do slopes decline through time or is the situation more complex with parallel retreat or time-independent landforms? Are meanders stage significant? Whether rivers meander or not, as has been explored earlier, is patently not simply a case of the 'age' of the landscape. This list is far from complete and the treatment of the cycle here has been concise

and selective. As this book is concerned with process and pattern it has not been a suitable framework for discussion. Many books give clear and detailed accounts but as you read them, however, it is important to ask two questions—what are the assumptions of the model, and does it now match our current knowledge of landscape processes?

Reconciling views may be possible however if the sporting phrase 'horses for courses' is kept in mind. In other words if the *scale* of enquiry is clarified. Over long time periods cyclic frameworks may be useful in aiding explanation of landscape. Over very short periods, several hours during a flood for example, steady-state views where none of the variables change with time may be more appropriate; similarly with respect to area. Continental-sized units may be appropriate for cyclic explanations whilst a slope segment of a few metres is more amenable to steady-state ideas.

After the somewhat broad-scale treatment of this section in the remainder of the chapter the focus will be over the shorter term with an examination of the relationships between some basin processes and man.

WATER AND MAN

Water, sediment, salt and man: the Colorado

For its size the Colorado is the most utilised and controlled river in the world. In the USA the lower Colorado generates $500 billion in farm sales a year, public water supply yields $57 million, electricity generation $21 million and recreation $250 million. With its water shared between seven American states and Mexico it is no wonder it has been argued over!

The regime of the river in its natural state may be determined from the Lee's Ferry records before 1963, when the upper part of the basin was relatively unaltered (Fig. 4.51B). The mean discharge rose to a pronounced peak in June, with lower discharges between September and May. The variations in monthly figures for individual years showed pronounced variations in the months from April to July. Why did the river behave in this way?

The simple pattern of total precipitation is only part of the story, the proportion of this which ends up as runoff is more significant. Figure 4.51C shows the distribution of runoff over the basin. Most of the area has less than 25 mm, the larger runoffs originating in the higher and northern parts of the basin. Here, in the cooler environments of the Rockies, Wind River Mountains and Uinta Mountains, evapotranspiration losses are less. Runoff itself reflects precipitation and evaporation over the basin, but the latter is higher in summer so another explanation has to be sought for the regime's summer peak—snowfall (Fig. 4.51D). As temperatures rise in spring the mountain snowpacks begin to melt, feeding water into the upper tributaries. Snowmelt thus explains the summer discharge peak. Variations in snowfall, together with the date and intensity of thaw, help explain the year-by-year variations revealed in Fig. 4.51B.

Later in the summer thunderstorms over the Colorado Plateau also contribute to discharge variations. They also play their part in generating the sediment load of red silt which gave the river its Spanish name. The semi-arid environment of the plateau with its partial vegetation cover, intense convectional storms and areas of Permian and Triassic sandstones provides an ideal environment for the production of high sediment concentrations in the streams.

During the last ten million years the river spread a delta across the upper arm of the Gulf of California. In the now isolated northern part of the Gulf intense evaporation in the hot arid climate occurred, leaving the Imperial Valley depression 72 m below sea-level. In 1901 the beginnings of irrigation and river control saw the diversion of Colorado water into the valley. In 1905 the Colorado burst its banks near Yuma and for two years it poured uncontrolled into the depression, forming a large lake. With no natural outlet, heavy evaporation over 70 years and saline water in-

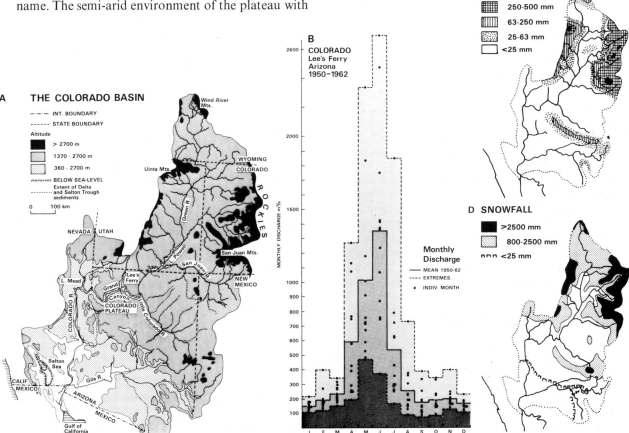

4.51 The Colorado basin

flows from drains in the irrigated area, has meant that the salinity of this lake (the Salton Sea) has steadily increased. It is now saltier than the ocean.

The Imperial Valley, with daily temperatures over 35°C from May until September and mild winters, is now the focus of a lucrative agricultural area. Farmers have been quoted as saying 'we'd just as soon not see rain at all'—it erodes seedbeds, washes off pesticides and muddy fields bog down vehicles. The water carried by the All American Canal typically costs $2·30 an acre foot (123 m³) and is used in the production of a vast range of crops. In terms of income, feedlots for fattening beef cattle are extremely important, a fact somewhat at odds with the 'winter salad bowl' image of the valley. Further to the north the Coachilla Valley has received Colorado water only since 1949, prior to that relying on local ground water. Being further away from the river its water costs are higher, typically $3·25 an acre foot. Since each acre can yield more than $1,300 a year the cost of water can be seen in perspective.

The nerve centre for distributing the water to this area is at the Imperial Dam, which diverts all but 4% of the Colorado from its natural channel. Farmers phone in their water orders and water is sent on its three-day journey to the furthest fields. The final stage of allocation is done by the 'zanjeros' or ditch riders who are responsible for opening and closing sluices on the final part of the distribution network.

The Hoover Dam is the lynchpin of river control on the lower Colorado. Completed in 1935 it can store in Lake Mead water equivalent to two years' flow of the river. Figure 4.52 shows the location of the dams which regulate flow and divert it for irrigation, drinking water and for power generation. Turning the unruly Colorado into virtually a series of lakes has created a new facet to the local tourist industry, worth ten times as much as the power generated. Lake Havasu City is one of the more familiar developments to English readers as the old London Bridge was re-erected there as part of its attractions.

In the lower Gila valley the Wellton–Mohawk irrigation district, begun in 1952, has been noted for its saline water problems. The Gila itself is naturally more saline than the Colorado, especially in terms of sodium, but the problem has concerned water which has been pumped from the ground to stop it rising to root level in the irrigated fields. This waste water was discharged into the Colorado just *upstream* from the

Morelos Dam, which the Mexicans use to divert water to the Mexicali and Yuma valleys.

This brings us to the problem of water allocation. In 1922 the Colorado River Compact divided the river's annual fifteen million acre feet equally between the upper and lower basin states. In 1944 the US promised Mexico one and a half million acre feet, but no mention was made of water quality. The allocation totalled sixteen and a half million acre feet. Unfortunately since the Compact the river's flow had declined to fourteen million acre feet, leaving a hypothetical deficit of two and a half million acre feet. As the upper basin states had not been using their share California was able to take more than her allocation. With the expansion of irrigation in the lower Gila there was less American surplus available to leach the salts from the Mexican land. The Wellton–Mohawk drain water was also saline, reaching 3,000 ppm, compared to the 920 ppm of lower Colorado water, which had been within the tolerance of crops. The situation was serious and the Americans built a by-pass for the Wellton–Mohawk water around the Morelos Dam in 1965. Unfortunately 40% of the Mexicans' irrigation is achieved by using wells to collect water from within the delta sediments. These delta aquifers were still being recharged by the drain water! At the same time Lake Powell was filling in the upper basin, reducing higher discharges in the lower basin which might have helped dilute this saline water.

Another dimension to this complex of allocation problems is the Central Arizona Project (CAP), due for completion in 1985. The rapid urban growth in the Phoenix and Tucson areas has led to massive demands for water. During the mid-1970's pumping from wells in these areas was taking out two million acre feet more water than is naturally recharging each year. This 'mining' of water has led to water-table lowering of up to three metres a year in some places. The CAP will take 1·2 million acre feet from the Colorado above the Parker Dam and transfer it south-eastwards. This obviously does not meet the area's deficit, but as the engineers say 'it buys the area time' to plan other ways to bring supply and demand into balance. Once the CAP is in operation California's allocation will fall but its search for, and development of, alternative resources are beyond our scope here.

Before summarising the lower Colorado situation the impact of man upstream in the *Grand Canyon* area will be examined. On and along the borders of the

Colorado Plateau the river and its tributaries have carved a spectacular network of canyons. This canyon cutting appears to have taken place in a series of pulses culminating in the Quaternary when climatic fluctuations caused alternate periods of downcutting and aggradation. In places the Grand Canyon is 24 km wide and more than 1,500 m deep. The river is now entrenched in the inner canyon flowing through Pre-Cambrian schists and granites. Above these stretch a sequence of sedimentaries dating from the Cambrian to the Permian and Triassic. The detailed profile of the Canyon with its cliffs, sloping benches and platforms mirrors the relative resistance of the limestones, shales and sandstones involved.

Between Lee's Ferry and Lake Mead, over a distance of 450 km, the river drops from 914 m to 259 m. Most of this fall is achieved in the 161 rapids which

4.53   The Hoover Dam and Lake Mead

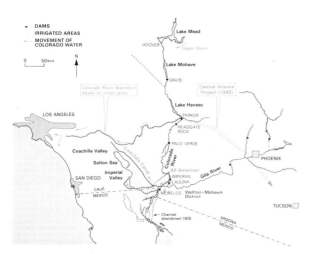

4.52   The lower Colorado

in total occupy only 10% of the channel's length. Contrary to simple ideas of a river attempting to remove such bed irregularities these rapids seem to have remained a part of the Colorado's quasi-equilibrium.

Discharge measurements show that floods exceeding 3,400 m³/s occurred every few years and the mean annual flood was 2,264 m³/s. How much sediment these carried depended on whether the runoff was from snowmelt or from thunderstorms over the plateau itself. The latter produced higher concentrations. On average over the year the Colorado carried 140,000,000 tons of material through the Canyon. The highest daily figure was the 27,600,000 tons carried on 13th September 1927 with a discharge of 3,538 m³/s. Imagine that tonnage being carried by more than 900,000 trucks and the immense capability of rivers as gutters of the landscape is all too apparent.

What were the geomorphological effects of such discharges? Figure 4.57 shows the scouring and re-filling of the river bed with the passage of the 1956 meltwaters, a typical change with floods in alluvial channels. Within the Canyon itself two sets of terraces were formed, one 9 m above the river corresponding to the infrequent 3,400 m³/s discharges and the other at 5½ m related to the 2,264 m³/s discharges. These ter-

4.54   The Grand Canyon

races were found especially at locations where river velocity was reduced. At the mouths of tributary canyons, fans of debris across the main canyon formed rapids, above and below these deposition occurred in the reverse eddies. These terraces appeared to be in quasi-equilibrium; floods with high-sediment concentrations building them and the low-sediment concentration floods causing a net removal of material.

This *delicately adjusted natural system*, together with its associated flora and fauna was disrupted by two events. The construction of the Glen Canyon Dam and the growth of recreation travel on the river.

The *effect of damming* can be seen dramatically in Fig. 4.58 which shows the sedimentation which has occurred in Lake Mead. In the case of Grand Canyon itself after the Glen Canyon Dam was finished in 1963 *sediment concentration* at Lee's Ferry (see Fig. 4.56) dropped by a factor of 200. At Phantom Ranch the reduction was only $3\frac{1}{2}$, a figure emphasising the role of tributaries in feeding sediment into the river. Such a reduction in sediment load means that the river's ability to erode and transport is increased. The dam also smooths out the *flood peaks* (the mean annual flood at Lee's Ferry has been reduced from 2,264 m³/s

to 764 m³/s) which means that the higher alluvial terraces mentioned above are no longer periodically reworked. Finally there is the fluctuation in discharge (and river level) reflecting the demand for electricity and the operation of the dam. Over a twenty-four-hour cycle demand varies by a factor of five which leads to variations of up to 4·6 m in river height. Discharge also drops at weekends and on public holidays when less power is generated, but rises to its yearly maximum when California's air conditioners are switched on during summer.

Such changes in the hydraulic regime and sedimentation following the completion of the Glen Canyon Dam have naturally affected the *alluvial morphology* and ecology of the Grand Canyon. These are summarised in Fig. 4.60. The higher terraces suffering wind removal of the sands and silts as well as vegetation encroachment, tamarisk, willow and arroweed creating a particularly dense barrier. The lower terraces are directly eroded by the river and also by seepages resulting from the daily fluctuations of water level. Before the dam was built during the autumn the Colorado typically lost competence and capacity as its discharge declined. It therefore deposited fresh material to replace that scoured away in the rising stages of the previous flood. A new equilibrium has to emerge in response to the changed hydraulic regime. Interestingly the effects of the dam are felt less as distance downstream increases and the contribution of tributaries is made. In the spring of 1973, for

4.55   The Inner Gorge of the Grand Canyon. The effects of the Glen Canyon Dam upstream and recreation pressures are discussed in the text and Fig. 4.60

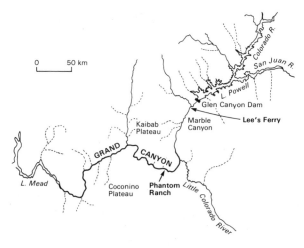

4.56   The Grand Canyon area

example, a 962 m³/s flood on the Little Colorado left
up to a metre of sand and silt as bars and terraces in
the Canyon. These were then reworked by the river,
illustrating that a new dynamic equilibrium is being
established in the Lower Canyon. Glen Canyon Dam
has a design life of 400 years, after which Lake Powell
will be sediment filled. The Colorado will then be faced
with re-establishing a new equilibrium.

In terms of *ecology* one of the vegetation changes
on the terraces has already been mentioned. In the
river itself fish adapted to the turbulent turbid water
of the natural Colorado, such as the squawfish, have
been replaced by clear-water species like trout. The

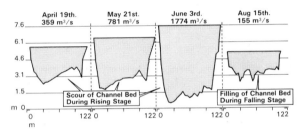

4.57 The Colorado, Lee's Ferry: channel changes during the
passage of the 1956 flood

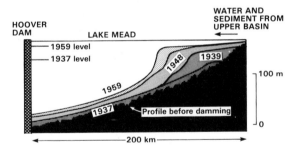

4.58 Sedimentation behind the Hoover Dam

clearer water has also allowed algal growth in shallows. Perhaps the most significant change, however,
has been man himself. It took until 1950 for the 200th
person to make the river trip. Since then the numbers
making this 'white water' trip and seeking the 'wilderness experience' has reached 14,000 a year. Camping
on the river bars during the 5- to 18-day trip (now having to travel so that rapids are passed at the high-water
pulses) means more than 40 people a day are using
the best sites. Trampling causes minor erosion and

4.59 Lake Powell above the Glen Canyon Dam

further environmental disturbance is caused by waste
left by visitors. Perhaps the biggest problem is sheer
numbers. People themselves destroy the 'wilderness
experience' they have come to enjoy. The Park Service
quota (set at 10,000 p.a.) is generally considered to be
at the upper end of the Canyon's carrying capacity for
this type of tourist.

The Colorado therefore illustrates some *general
principles of river behaviour*—the ways in which discharge and sediment are influenced by elements of the

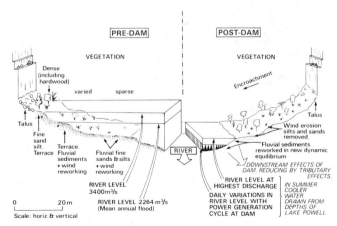

4.60 The effects of Glen Canyon Dam on the alluvial morphology and ecology of the Grand Canyon

basin environment. It also has the kind of water *management and allocation problems* which few other basins have yet faced. Unlike many natural systems the economy of the basin has prospered on positive feedbacks, or cumulative growth. The successful development of agribusiness (based on irrigation), recreation, power generation and urbanisation have fed their own further development. The 'image' of the south-west (its climate, scenery and growth) has aided its success. The problem is that all this has demanded water, which is a finite resource.

There are three facets to the water situation: supply, demand and quality. Demand, for example, could be reduced by even more efficient water allocation, by increasing its price and by reducing irrigation acreages. Saline drain water could be tackled by desalination. Supply could be increased by a range of techniques from the prosaic lining of ditches (*i.e.* cutting losses), to tapping further ground water, through to the dramatic proposals for cloud-seeding to increase winter snowfalls or massive inter-basin transfers of water from the northern US and even Canada. Such technological responses meet with growing resistance from environmentalists. The future will be determined by the shifting power balance between the promoters of growth, environmentalists, the state governments and the agencies of the Federal Government.

In conclusion, the words of J. Ives, an early explorer of the south-west. 'It seems intended by nature that the Colorado River, along the great portion of its lonely and majestic way, shall be forever unvisited and undisturbed', have proved rather unprophetic!

The Wye Basin

The Wye basin, in contrast to the Colorado, is considerably smaller and lacks the same environmental extremes. It does, however, neatly encapsulate some water management issues in the British context.

Figure 4.61 shows the basin together with rainfall and runoff for eleven sub-catchments. Rainfall varies across the basin, with larger totals in the west and at higher elevations. In this it reflects the origins of most British rainfall—frontal with westerly airflows and a substantial relief component. The difference between rainfall and runoff is an indicator of evapotranspiration losses. If you look at the figures you will see that the loss varies from less than 30% of rainfall in the higher and western areas to more than 50% in the east.

Runoff, the water available for streamflows, therefore is higher in the west. *The spatial pattern of input, loss and output* is typical of Britain as a whole.

How these inputs and outputs vary through the year is shown for the upper Wye and the Lugg in Fig. 4.62. Both show the normal winter peak in rainfall, a reflection of depression frequency. Runoff is also greater in winter but higher temperatures and plant growth during summer accentuate losses in that season.

In a nutshell this is the British surface water supply situation. Supply is greatest in winter, on higher ground and to the west. On the whole this isn't *where* water is required or *when* it is used most. Runoff variations in the British Isles, with exceptions like the drought mentioned on pages 36–7, have not been extreme. Water consumption, however, rises with economic growth, and supplies over parts of lowland, urbanised and industrialised Britain will need to be augmented. Figure 4.63 shows one proposed water resources strategy to cope with the patterns of supply and demand. With this proposal a series of strategic reservoirs in Wales would be used to regulate river flows. From these rivers water could be taken to areas of demand or transferred to other basins.

One of the dramatic facets of river behaviour which has a direct impact on man is *flooding*. The largest

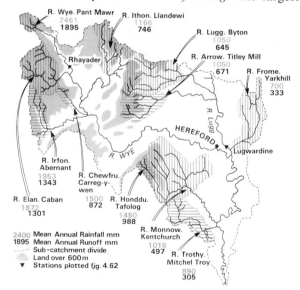

4.61 Sub-catchments within the Wye basin: rainfall and runoff

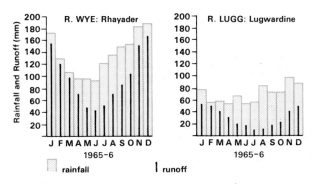

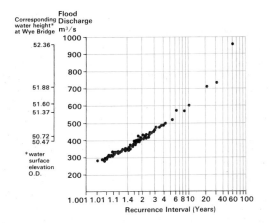

4.62 Mean monthly rainfall and runoff; the Wye and Lugg

4.65 Flood frequency of the Wye at Hereford 1908–68

recorded flood on the River Wye at Hereford occurred on 4th December 1960 and its extent is shown in Fig. 4.64. By a process of trial and error, or more realistically by building things and having them washed away, earlier societies adjusted to the *flood hazard*—using flood plain land as water meadows, for example, and by building sturdier and higher bridges. These adjustments were probably made to flood events of the sort of frequency which occurred in each generation's lifetime. It is interesting to speculate that the advantages of living and working close to the nodal bridge site (A in Fig. 4.64) were sufficient to offset the damage and inconvenience of floods. With more complex urbanised and industrialised societies adjustments cannot be left to chance or hard-won experience.

Magnitude frequency analysis is used to calculate the probability of a flood event happening. This is done by noting the magnitude of the largest flood each year. These are then arranged in rank order, 1 for the largest and so on, for the period of years for which records exist. The recurrence interval for each flood discharge is then calculated, $T = (n + 1) \div m$ where T is the recurrence interval, n is the number of years of observation and m the rank. Data for the Wye is plotted in Fig. 4.65. The horizontal axis is the Gumbell–Powell scale. (For a discussion of the Theory of Extreme Values see Newson in the further readings at the end of the chapter.)

When people's ground floors are flooded or their journey to work upset, the cry, ' "they" ought to do

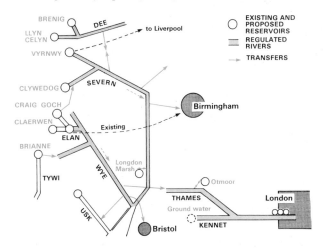

4.63 Part of a proposed national water grid

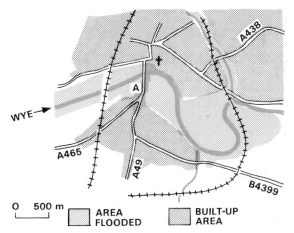

4.64 The Wye flood of 4th December 1960 at Hereford

something about it' is heard. Engineers then produce protection schemes involving embankments, raising bridges and so on. Unfortunately by the time the plans come before the decision-makers in local government, the ratepayers have forgotten the disruption of the last flood and resent having to foot the bill! In other words it is suggested that people's *perception* of the hazard diminishes with the passing of time.

Protection against extreme events is expensive, yet no protection involves, at the least, property damage, and at the worst, loss of life. The engineers' response in consequence is a compromise. Urban flood designs often revolve around the '100 year flood' whilst storm-water drainage from streets around the '10 year' event. The magnitude frequency analysis shown in Fig. 4.65 is used. Long-term hydrological records are rare; extrapolating outwards from a short run of records

is subject to error. Ignoring the 1960 flood in the case of Fig. 4.65 (some 958 m³/s) would have placed the 60 year flood at about 825 m³/s, as you can see by carrying the line of the smaller floods outward. It should be emphasised that the kind of approach shown in Fig. 4.65 is one of statistical probability, or prediction, not of forecasting. It doesn't tell us the day and time of the '250 year flood'.

Probably less dramatic and newsworthy is the problem of rural flooding. Figure 4.66 shows the lower Lugg valley. Great stress has been placed on 'normal' rivers in this chapter—unfortunately the Lugg has a grossly inadequate channel capacity. It is able to contain only 23% of the water involved in the 'mean annual flood'. The consequences of this and the resultant flood-protection scheme are shown in Fig. 4.66.

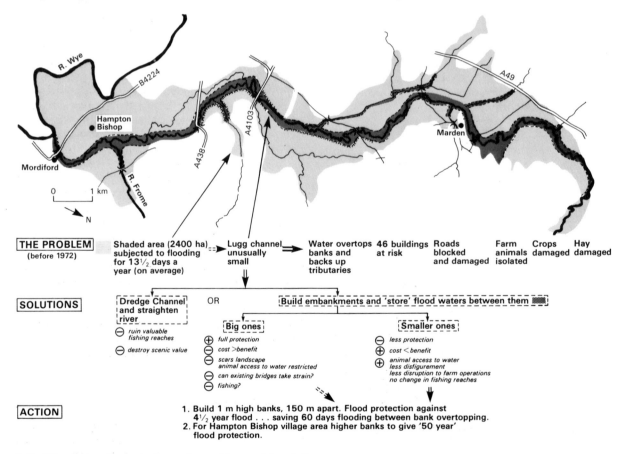

4.66   Flooding on the lower Lugg: the problem and its solution

The problems of low flow: the Ganges

In contrast to the Wye and the urban and rural flood hazards reviewed in the last section, many areas of the world are beset by the problems of low and erratic flow. The regime of rivers in the Indian subcontinent mirrors the seasonal rainfall patterns associated with the monsoon and its variability from year to year (Chapter

2). This is dramatically seen in the monthly discharge of the Ganges, plotted over a period of eleven years, shown in Fig. 4.67. The kind of problems this poses for agriculture, the well-being of rural society and economic development in general is well covered in other sources. Our focus here is on the allocation of water at times of low flow between India and Bangladesh.

The critical period is in April when minimum daily discharges at Farakka may fall to 1,100 m³/s, this from a drainage area of 951,600 km²! India was faced with two problems, the economic development of the area and Calcutta. The Hooghly River connects Calcutta to the sea. It is no longer a distributary of the Ganges and has been subject to silting and shallowing. One way to solve this access problem for the port would be to increase the flow in the Hooghly. This would also have the advantage of improving the water supply situation for the city. For these and other reasons India built a barrage at Farakka, 20 km from the Bangladesh border and in 1975 began diverting water southwards. At times the diversion amounted to 73% of the flow.

Bangladesh, left with as little as 27% of the flow, claimed that this was inadequate for the third of its population living in the downstream area. Irrigation was hindered, river transport impaired and fishing and industry affected. The low flows also increased health hazards and the intrusion of sea water into the delta.

Conflicts over water use are seldom simple or easily solved. However in November 1977 Bangladesh and India signed an agreement lasting for five years concerning the allocation of water in the latter part of April. In the agreement India would take 580 m³/s and Bangladesh 976 m³/s. The agreement also included provision for a study to be made of ways of increasing the flow in the critical period before the arrival of the monsoon. You might like to consider the possibilities. The two logical alternatives are storing the monsoon rains or diverting water. Where could this be done? Imagine the problems of the capital required for the engineering works themselves and, of course, the myriad adjustments which would have to be made in the workings of society.

Saskatchewan

Our final case study, from the Prairie Provinces of Canada, is an example of multi-purpose water management. The area is shown in Fig. 4.68. Surface

R. Ganga (Ganges), Farakka

(Drainage area 951,600 km²)

DISCHARGE

MEAN (11 year)

HIGHEST

LOWEST

4.67 The River Ganga (Ganges) at Farakka

water supplies here fall into two categories. Firstly water from the west, originating from precipitation which has fallen on the Rockies and their foothills. This is carried eastwards towards Hudson's Bay by the North and South Saskatchewan Rivers. The regime of the South Saskatchewan is shown in Fig. 4.69A and 90% of its water originates in the foothills and mountains of western Alberta. In view of what has been said earlier about the Colorado the reasons for the seasonal pattern and its variability should be obvious.

The other source of water is the local prairie streams themselves. The environment shown in Fig. 4.70 and the climate depicted in Fig. 4.69C result in a discharge pattern shown in Fig. 4.69B. Prairie streams therefore have small, seasonable and unreliable runoffs. Figure 4.68 also shows large areas of internal drainage. Here runoff, when it occurs, collects in shallow depressions (sloughs) or larger lakes. From these it evaporates, leaving saline deposits in the sloughs, or percolates as ground water recharge. Such supplies being small in amount, unreliable in quantity and poor in water quality might be appreciated by wildfowl but present problems in management.

Stated very concisely, that is the supply side of the equation—what of demand? This has a number of facets each having its own distribution and annual pattern which will have to be integrated in any management scheme. Firstly there is public water supply. In 1960 only 45% of the towns had a public water supply and during the previous decade urban population had increased by 42% to 527,484. Population distribution

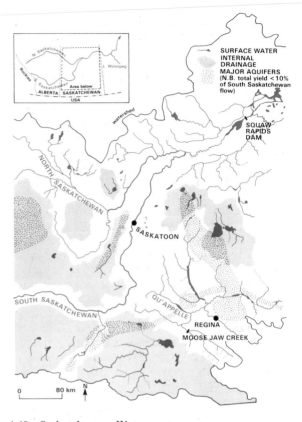

4.68   Saskatchewan: Water resources

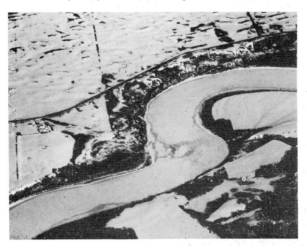

4.70   The ice-covered South Saskatchewan River in winter

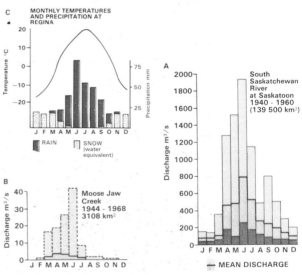

4.69   River discharge and climate

is shown in Fig. 4.71. Secondly there was a shortage of water recreation sites, especially in the Saskatoon area. Thirdly there was the question of increasing electric power demands. Thermal generation provided base load but the ability of the Squaw Rapids hydro site to offer peak load 'top-ups' was hindered by the low winter flows of the Saskatchewan River. Fourthly although endowed with a short growing season, and distant from markets, there was some potential for irrigation in the province to widen the crop range and provide more fodder. Fifthly there was increasing water demand associated with economic development and industry. One component of this was the anticipated growth of potash mining which would need an additional 369,000,000 m³ of water a year.

These facets of demand apparent in the late 1950's and 1960's all had their own distribution and the maps are designed to illustrate this component of demand. Some of these uses have even demands during the year, others peak in summer or in winter. Bearing in mind these questions of *where* and *when*, you might like to think of the kind of allocation strategy required. Figure 4.72 shows the general terrain of the area. In the light of what has been said above and the information contained in the figures, where would you place

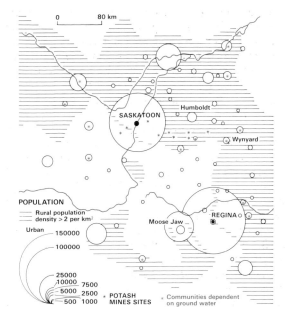

4.71   Saskatchewan: water demands

a multi-purpose scheme and how and when would you allocate water to the various uses?

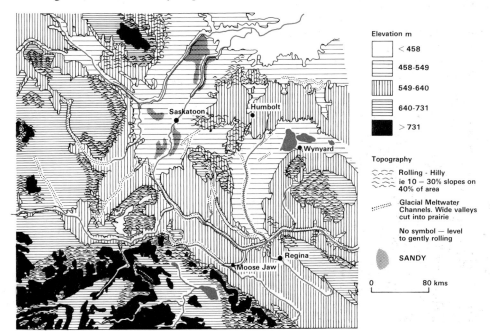

4.72   Southern Saskatchewan: relief and terrain

The South Saskatchewan Water Development Projects were designed to meet the supply and demand situation outlined above. Construction started in 1958 and the completed scheme is shown in Fig. 4.73. It may well reflect your own ideas on an appropriate water resources strategy. The South Saskatchewan River is impounded by the Gardiner Dam and as a result of the topography a further retaining dam was built on the Qu'Appelle Arm. Figure 4.73 shows the general movement of water and the use which has been made of the coulées and existing rivers. The annual operation of the Diefenbaker Lake and the allocation of water to the various uses is shown on the right of the map. This is intended to reinforce the points made earlier about the need to consider precisely when supply and demand exists in any multi-purpose scheme.

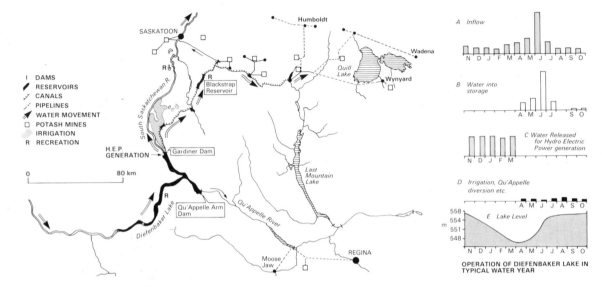

4.73   The South Saskatchewan Water Development Projects

4.74   The 5 km long Gardiner Dam. The Qu'Appelle Arm
Dam lies at the top left (see Fig. 4.73)

CONCLUSION

The focus of this chapter has been *water on the land*. Its significance to man as a *resource* has been explored in the preceding pages and from time to time opportunities were taken to re-examine the erosional, transportational and depositional work of water. This last section also focused on river regimes and runoff variations in time and space. The various pathways of water discussed in the first part of the chapter underpins these variations.

Water on the land is at the interface between the atmosphere and lithosphere. Much of the material introduced in earlier chapters is therefore pertinent to an examination of the drainage basin. Only a selection of these concerning structures, lithologies, weathering, slope processes and precipitation were taken up. Many other inter-relationships have been left unexploited here and the further readings should be explored if you wish to take this further. Most importantly, however, the role of vegetation has received little direct attention although reference has been made to the significance of plants in connection with various soil and ground water processes. The focus of the next chapter will therefore be on the role of organisms.

Review Questions and Exercises

1. Describe the various pathways of water within the basin hydrological cycle.
2. Why do rivers flood and what are the consequences of this?
3. Describe and justify how you might investigate the input and output from a small (*i.e.* 20 km²) catchment.
4. Figure 4.21 showed different sediment concentrations in different seasons. How do variations in base flow, soil moisture and precipitation character produce these differences?
5. What are the geological influences on drainage patterns and river runoff?
6. Describe what you would expect to happen to five pebbles (2 cm in diameter), 50 sand grains and 50 clay mineral particles found on a slope 500 m from a stream channel, before the material ultimately arrives in the sea. They are at an elevation of 1,500 m some 300 km from the sea. Would there be any differences in their life histories under different climatic conditions?
7. Describe the operation of the South Saskatchewan Water Development Projects and explain what dimensions of the physical and socio-economic environment it reflects.
8. Describe and account for the various landforms produced by river deposition.
9. Explain the reasons for the differences in the water balances shown in Figs. 4.15 and 4.16.
10. Figure 4.75A shows an analysis of flood magnitude in England and Wales. The inset graph indicates how the six areas were defined. The data concerns the mean annual flood which has a recurrence interval of 2·33 years. Figure 4.75B shows the maximum one-day rainfall for a two-year recurrence interval. How closely do these two patterns resemble each other? Can you find any evidence of a geological influence on the flood magnitude distribution and suggest how the basin hydrological cycle may operate in those areas?

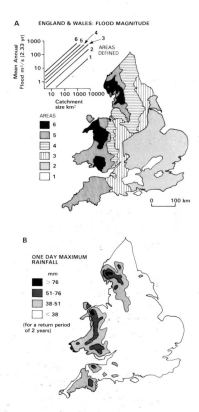

4.75 Floods in England and Wales

11. Describe how river channel processes and slope processes combine to produce river valleys.
12. As a physical geographer you have been asked by an international boundary comission to comment on the feasibility of using 'natural' and 'enduring' features such as divides and stream channels as boundary lines. What advice would you offer for the area shown in Fig. 4.76?

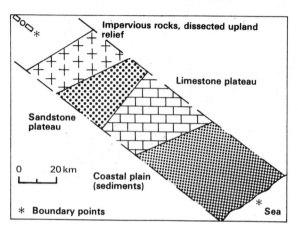

4.76   A boundary problem

13. Figure 4.77 shows the relationships between annual precipitation and runoff (A) and annual precipitation and sediment yields (B) for a number of sites in the hill land of the upper coastal plain in Mississippi. The sites have four types of vegetation. How would you describe these relationships and can you suggest reasons for the differences between the various land uses?

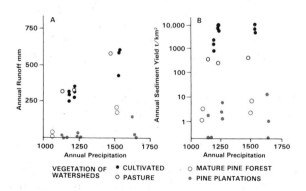

4.77   Precipitation, runoff and sediment yields

14. Look at Fig. 4.32, produce a sketch of it and indicate the various landforms. How have these been produced and what kind of changes would you expect during a flood on the one hand, and over a period of about 10,000 years on the other?
15. Figure 4.78 shows the distribution of annual runoff in the USA. Using atlas maps and other reference sources describe this pattern and relate it to relief, airflows, precipitation and temperatures.

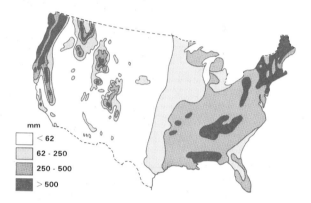

4.78   USA: annual runoff

16. Figure 4.79 shows the distribution of annual sediment concentrations in American streams. Which parts of the country have the highest and lowest concentrations? Can you suggest reasons for this pattern bearing in mind precipitation and vegetation/land use?

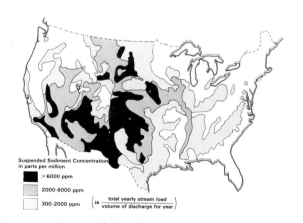

4.79   USA: suspended sediment concentrations

	Year	J	F	M	A	M	J	J	A	S	O	N	D	Year	MAX (date)	MIN
R. NIGER KOULIKORO (120,000 km²)	1950	286	145	74	42	65	128	730	2518	5198	5547	2391	768	1497	6440 30.IX	34
	1951	374	207	142	87	221	544	1583	3753	5349	5431	5433	1814	2085	6420 17.XI	69
	1952	751	419	208	116	100	186	1208	3221	5134	5513	2316	906	1677	6280 21.IX	78
	1953	508	238	143	81	97	560	2174	4436	6575	5152	2373	1069	1959	6960 10.IX	60
	1954	608	325	195	188	208	642	1947	4284	6127	5148	3280	1674	2061	6480 21.IX	130
	1955	760	432	289	207	222	677	2037	4076	6095	5985	2763	1286	2064	7456 29.IX	137
	1956	659	377	238	174	127	198	980	2272	4807	4561	1733	769	1410	6210 30.IX	107
	1957	394	186	108	56	67	315	1278	3674	6237	6678	3411	1192	1975	7247 21.IX	56
	1958	613	360	160	133	257	785	1411	2092	4305	4263	2132	1326	1492	5437 9.X	105
	1959	598	317	168	78	92	311	1345	2895	5556	4328	1770	740	1522	6946 28.IX	64
	1960	353	178	82	56	80	290	1354	3637	5789	4748	2122	819	1629	6550 30.IX	36
	1961	357	170	76	30	67	93	924	2877	5256	3209	1239	478	1235	6172 18.IX	25
	1962	204	89	37	20	115	202	1068	3203	6943	5895	2691	1132	1807	7798 25.IX	20
	1963	484	259	170	65	129	130	632	2351	4706	5686	2691	842	1518	7228 23.X	52
	1964	467	166	69	37	36	397	1150	3574	5277	4960	1744	890	1568	6640 6.X	22
R. NIGER DIRE (340,000 km²)	1950	1487	1096	686	207	85	56	74	718	1494	1950	2188	2370	1034	2405 25.XII	—
	1951	2222	1684	1018	372	87	129	423	971	1588	1946	2211	2430	1255	2557 144*	(55)
	1952	2534	2279	1717	1002	351	86	177	921	1688	2070	2284	2474	1463	2535 2.1*	65
	1953	2431	1981	1360	684	167	79	571	1338	1767	2114	2366	2551	1449	2595 26.XII	71
	1954	2455	1985	1359	696	238	169	593	1236	1769	2102	2363	2570	1459	2647 5.1*	111
	1955	2582	2242	1677	980	431	195	640	1319	1796	2138	2391	2589	1579	2640 7.1*	143
	1956	2550	2106	1442	799	276	89	186	786	1433	1870	2136	2276	1327	2300 24.XII	68
	1957	1931	1384	781	252	66	50	293	1032	1684	2034	2285	2519	1192	2632 11.1*	(50)
	1958	2579	2211	1647	900	331	230	675	1222	1648	1986	2207	2381	1498	2411 15.XII	127
	1959	2225	1791	1267	608	140	(50)	241	912	1546	1950	2197	2315	1268	2359 23.XII	44
	1960	1958	1438	786	237	52	(38)	300	1036	1620	1980	2200	2317	1164	2392 17.XII	35
	1961	2055	1562	944	342	64	(40)	845	1549	1905	2148	2098		1135	2269 9.XII	(23)
	1962	1690	1121	517	136	(40)	(30)	185	850	1546	1930	2206	2391	1056	2431 24.XII	(25)
	1963	2186	1762	1142	513	(110)	(60)	(87)	634	1357	1767	2033	2203	1151	2262 24.XII	35
	1964	1897	1464	820	238	(63)	(37)	324	958	1620	2022	2258	2343	1170	2418 18.XII	36
R. KOLYMA SREDNE-KOLYMSK (361,000 km²)	1940	114	83.0	71.0	52.5	2900	12600	5530	3990	5080	727	220	176	2630	16200 8.VI	42
	1941	102	62.0	57.5	52.5	222	12100	7390	4050	4320	2100	357	251	2590	22900 5.VI	50
	1942	122	79.0	59.0	46.5	64.5	15200	5420	5630	3510	797	287	200	2620	25100 4.VI	45
	1943	160	101	77.0	71.5	4470	8890	4120	4080	4500	2570	517	297	2490	16400 30.V	67
	1944	165	93.5	75.0	64.5	5950	6100	4350	4470	4380	2930	483	378	2460	14700 21.V	63
	1945	176	95.5	67.0	58.0	5520	6070	3790	2880		1010	249	164	2340	23400 23.V	55
	1946	108	64.5	58.5	51.0	2640	8380	5980	3570	3860	1340	351	192	2220	12400 31.V;1.VI	47
	1947	114	76.0	58.5	52.0	71.0	13100	9270	4540	2440	823	263	175	2580	19400 25.VI	49
	1948	130	83.5	57.0	50.5	4060	5050	3030	3890	1930	366	183	142	1580	15800 17.V	48
	1949	83.0	60.0	54.5	57.0	1710	7620	2740	2700	1950	332	177	132	1470	15300 30.V	51
	1950	84.0	58.0	44.0	34.5	112	13500	9230	9890	4340	494	263	223	3190	19200 22.VI	30
	1951	124	87.0	70.5	60.0	2650	16000	5020	3320	1540	675	279	186	2500	19100 24,25.VI	52
	1952	100	70.0	58.0	52.5	736	10800	6210	3110	2360	475	262	178	2030	20900 5.VI	50
	1953	101	69.0	57.5	48.0	5570	8670	5440	4960	3960	931	433	292	2540	20800 28.V	45
	1954	129	80.5	58.0	52.5	442	14200	6580	3170	4870	1100	368	228	2610	16500 19.VI	47
	1955	110	67.5	63.0	49.0	1900	8080	5230	3420	1770	345	233	197	1790	23400 30.V	43
	1956	135	104	77.0	70.0	2890	17300	6450	2580	2610	434	261	186	2760	20000 25,26.VI	66
	1957	106	74.0	65.5	58.0	1850	9900	4000	4560	1580	740	291	188	1950	15600 7,8.VI	55

17. Using the data in Fig. 4.80 plot the discharges and monthly means for these rivers (using the method shown in Figs. 4.67, 4.68, etc.). Using reference texts suggest reasons for the discharge patterns and comment on what kind of water resource management strategies might be required.

4.80   River discharges

Further Reading

Chorley. R. J. (Ed.), 1969, *Water, Earth and Man*, Methuen.

Dury, G. H. (Ed.), 1970, *Rivers and River Terraces*, (Geographical Readings), Macmillan.

Gregory, K. J., & Walling, D. E. 1976, *Drainage Basin Form and Process*, Arnold.

Leopold, Wolman & Miller, 1964, *Fluvial Processes in Geomorphology*, W. H. Freeman.

Morisawa, M., 1968, *Streams*, McGraw Hill.

Newson, M. D., 1975, *Flooding and the Flood Hazard in the United Kingdom*, Oxford University Press.

Smith, D. I., & Stopp, P., 1978, *The River Basin: An Introduction to the Study of Hydrology*, Cambridge University Press.

Strahler, A. N., 1975, *Physical Geography*, Wiley.

Ward, R., 1978, *Floods*, Macmillan.

Weyman, D., 1975, *Runoff Processes and Streamflow Modelling*, Oxford University Press.

Weyman, D., & V., 1977, *Landscape Processes*, Allen & Unwin.

Fieldwork

Field techniques and projects are described in—

Anderson, E. W., 'Drainage Basin Instrumentation in Fieldwork', Geog. Assoc. Teaching Geography, Occasional Paper **21**.

Briggs, K., 1977, *Sediments*, Butterworth.

Gregory, K. J. and Walling D. E. *Field Measurements in the Drainage Basin*, Geography, 54, 4, 277–92.

Hanwell, J., & Newson, M. D., 1973, *Techniques in Physical Geography*, Macmillan.

Weyman, D., & Wilson, C., 1975, 'Hydrology for Schools', Geog. Assoc. Teaching Geography, Occasional Paper **25**.

Sources of Streamflow Information

H.M.S.O., *Surface Water Year Books* and the publications of the Water Resources Board. Regional Water Authorities and Drainage Boards annual reports.

U.N.E.S.C.O., 'Discharge of Selected Rivers of the World', Paris, 1972, and publications connected with the I.H.D.

5 The Living Earth

INTRODUCTION

Our concern in this chapter is with the various life support systems of spaceship earth. The living envelope of the planet is known as *the biosphere*, and consists of the oceans and those parts of the land surface where moisture and temperature conditions are suitable for life. Within the biosphere are a host of constituent ecosystems some of which will be examined in this chapter. An *ecosystem* comprises all the communities in an area together with its non-living environment. It is therefore a functioning unit, or system, of plants, animals, soils and atmosphere. Interactions between vegetation, animals, terrain, climate and man will be examined in a number of environments. The central ideas of energy flow and material cycling will be initially introduced in the context of two contrasting environments, the northern seas and the tropical rain forest. The interactions of organisms and the formation of soils will then be explored before we return to the functioning of the biosphere as a whole.

AN INTRODUCTION TO ENERGY FLOW AND MATERIAL CYCLING: THE NORTHERN MARINE ECOSYSTEM

All life depends on a single source of energy, the sun. Solar energy is held briefly in the biosphere before being re-radiated as heat, so the continued functioning of the biosphere and its ecosystems depends upon this continual flow of energy. This solar energy enters the biosphere through the photosynthetic production of organic matter. The process of photosynthesis is thus fundamental for maintaining life on earth. In this process chlorophyll-bearing organisms, using the energy of the sun, convert CO_2 and water into carbohydrates, releasing O_2 as waste. Such organisms are known as *primary producers*.

In the ocean primary producers are the simple single-celled green plants of the sea, the phytoplankton. Since they depend on sunlight they can only exist in the upper, illuminated part of the sea. The actual process of photosynthesis is relatively inefficient, the proportion of solar energy converted into living systems being usually less than 1%. In the case of our phytoplankton, as with all primary producers, a considerable proportion of this is actually used in the functioning of the organism and the amount of energy 'locked up' as organic matter is even smaller.

This material, the net production of the primary producers or *first trophic level*, forms the food supply for the *second trophic level* or herbivores. In the sea these consist of various species of zooplankton, small organisms which 'graze' on the phytoplankton. The herbivores in turn use almost all their energy input and only a tiny fraction is locked up as organic matter.

The *third trophic level* consists of carnivores who eat the flesh of the herbivores: fish, seals and whales in the case of the northern seas. These in turn may be preyed on by a *fourth trophic level* of carnivores such as polar bears and man. Again the actual conversion of energy into organic matter is 'inefficient' and the majority of the food energy input is lost with the normal functioning of the fish, seal or whale.

The flow of energy through the various trophic levels with the losses at each stage is shown diagrammatically in Fig. 5.1. The linear sequence from marauding polar bear, through unfortunate seal and shellfish to the phytoplankton is known as a *food chain*. In more complex environments with a greater number of species at the various trophic levels the

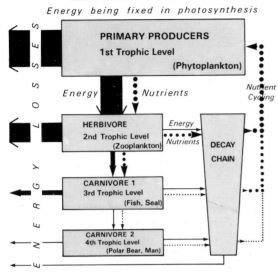

5.1 The northern marine ecosystem: energy flow and material cycling

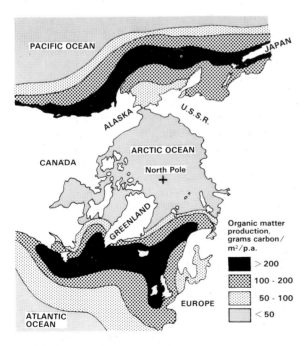

5.2 Marine organic matter producton

whole network of feeding habits is known as a food web.

Figure 5.2 shows the circumpolar distribution of organic matter production. In general terms *primary production is controlled by five factors*. Fundamental is the available *light*. This will obviously vary with latitude and decline towards the pole. The distribution shown in Fig. 5.2 doesn't really conform to this as parts of the Atlantic and Pacific are as low in productivity as the Arctic Ocean itself. A second factor is low *temperature*. This may be significant on land but temperature variations within the sea are far from severe and organisms have easily evolved to cope with the near-freezing-point waters. The third general factor affecting productivity is *water*, obviously a non-starter with this example as water is liberally available in the sea! (It is, however, important when considering terrestrial productivities and as some general points are being made here it has been mentioned for completeness.) The explanation for the variation in productivity, shown in Fig. 5.2, must therefore lie with the remaining two factors, *nutrient supply* and the ability of the ecosystem to use and *circulate* materials.

In Fig. 5.1 and the earlier discussion the flow of energy from phytoplankton to polar bear was established. It is important to realise however that energy may also flow into the decay chain (shown on the right of the diagram). Dead phytoplankton, zooplankton, fish and seals are fed upon by a range of decomposers. In this process these detritus feeders return nutrients which the primary producers had originally removed. Figure 5.1 (broken lines) shows this cycling of nutrients. Nature supplies a constant flow of solar energy but provides no new input of material which must therefore be *recycled*.

All ecosystems are characterised by these two processes of energy flow and material cycling. The material recycled consists primarily of oxygen, carbon and hydrogen (the constituents of the carbohydrates synthesised during photosynthesis) together with elements such as nitrogen, phosphorus, calcium, sulphur, etc. These are the so-called biogeochemical cycles, similar to the hydrological cycle in that they consist of various stores and transfers. Some of them operate over decades, some over hundreds of millions of years, but they are all vital to the continued existence of the biosphere.

The cycling of material may therefore be the most satisfactory explanation for the variations in produc-

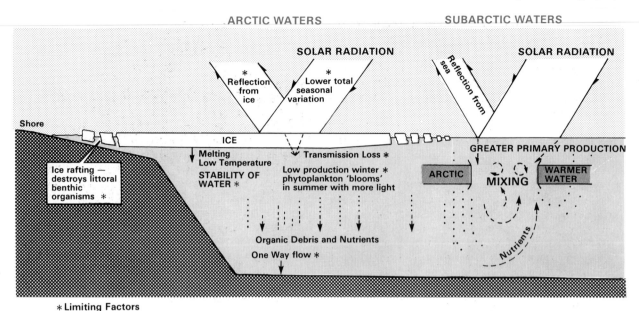

* Limiting Factors

5.3   Productivity controls in arctic and subarctic waters

tivity shown in Fig. 5.2. In the marine environment when organisms die their remains fall towards the sea bed. Under the polar pack ice the various layers of sea water are intensely stable and little mixing occurs so that the products of decay are not returned to the upper illuminated layers. In the 'subarctic' waters, on the other hand, intense mixing occurs, nutrients are returned to the surface layers and a higher primary production exists. Figure 5.3 summarises this point and also indicates how light and temperature influence the rate of primary production.

THE TROPICAL WORLD

The rain forest

Having established the ideas of energy flow and material cycling in one of the simplest ecosystems we now turn our attention to the most complex—the tropical rain forest. Within it there are three formations, the American, African and Indo-Malaysian and their distribution is shown in Fig. 5.4. These areas are characterised by uniformly high temperatures, with annual means between 20° and 28°C and annual ranges

of as little as 2°C. Diurnal temperature ranges exceed the annual variation, ranging between 3° and 16°C. Rainfall everywhere exceeds 2,000 mm a year and is well distributed through the year with no month having less than 60 mm.

The tropical rain forest is an association of producing, consuming and decomposing organisms and in common with all ecosystems it derives its ultimate energy from the sun. More than 90% of the *biomass* (the weight of organic matter per unit of area) consists of evergreen trees. They have straight slender trunks, usually thin smooth bark and frequently buttress-like flanges on the lowest three metres of trunk. Their leaves, which in many cases are cast off and regrown continuously with no annual rhythm, are simple in shape. They are often 'leathery' in appearance and this cutinisation enables the leaf to remain rigid in the mid-

5.4   The extent of the tropical rain forest

day heat and thus continue the photosynthesis process and the transpiration of water brought up from the roots. Drip tips, or extensions at the points of the leaf, allow moisture from the frequent rains to be shed quickly so that photosynthesis and transpiration can continue almost continously. Richness of species, or floristic diversity, is a feature of the rain forest with as many as 200 tree species per hectare. Individual trees of the same species may therefore be quite widely spaced.

In addition to the mature trees and their seedlings there are only a few shrubs and herbs in the deeply shaded lower layers. Woody climbing plants (lianes) and plants using trees for support to reach the light (epiphytes) are common. There are also many parasitic plants (those drawing nutrients from their living host) and saprophytic plants (those feeding on dead organic matter).

A characteristic feature of the forest is its *stratification*. The existence of three interlocking layers is widely recognised although it may not be that apparent from a purely visual examination (Fig. 5.5). This layering is shown in Fig. 5.6 and consists of the canopy layer 20–30 m high above which are the emergents. Beneath the canopy is a layer of trees with long tapering crowns, about half of these are usually saplings of the larger trees. Beneath this layer is a thin scattering of ferns, herbs and seedlings but a considerable proportion of bare ground. The myth of impenetrable jungle dies hard, perhaps the early 'explorers' liked to over-dramatise their exploits! Only when light reaches the surface does dense vegetation exist, along river banks, in man-made clearings or where a forest tree has fallen, for example.

The floristic diversity of the forest is paralleled by faunistic diversity. In the various layers of the forest there are different foods, opportunities for concealment and possible ways of movement. In other words a wide variety of *ecological niches* to be exploited. In the canopy layer, for example, we may find flying squirrels, monkeys, butterflies, frogs and hosts of insect species which never reach the lower layers. On the trunks we may find martens and baboons and, on the ground, tapirs, anteaters, deer, ants, termites and so on. Overall there is a low ratio of animal to plant life. In the Amazon rain forest, for example, the total tree biomass is 900 tonnes/ha and the total animal biomass 0·2 tonnes/ha.

In the rain forest we therefore have sufficient

5.5 Tropical rain forest structure revealed by road construction in the Amazon Basin

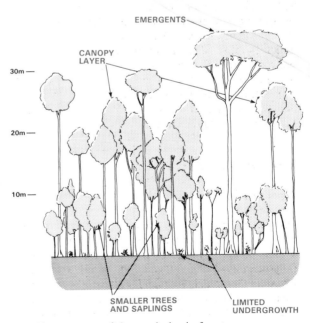

5.6 The structure of the tropical rain forest

warmth and moisture to permit the year-round growth of a diversity of plants and the production of a large plant biomass. This luxuriant vegetation, however, exists on poor soils. How can we explain this paradox?

The answer lies in considering two of the elements affecting ecosystem productivity which were introduced in the discussion of the northern marine ecosystem, *nutrient supply* and *material cycling*. Nutrient supply for terrestrial ecosystems is derived from the soil and from rainwater. In the tropical situation weathering processes have been proceeding for so long and with such vigour that the weathering front (where appreciable amounts of fresh minerals are liberated from the rock) is often deep below the surface and unavailable for plants. The frequent rainfall also removes soluble minerals being released in the weathering zone. Most of the nutrients, nitrogen, calcium, phosphorus, potassium and so on, are in fact stored in the living tissues of the organisms themselves and not in the soil. Dead leaves and branches together with the excretions and dead bodies of organisms fall to the ground surface. This continual supply of organic matter is broken down by the decomposer organisms such as fungi and bacteria. The minerals released are quickly absorbed by the roots of the growing plants within the upper few metres of the ground. The small nutrient stock of the rain forest is therefore quickly recycled and with no significant seasonal variations in climate this process is continuous. The small amount of nutrients washed downwards are easily replaced by those contained in the rainwater. In the Amazon Basin, for example, rain brings 10 kg of nitrogen, 3·6 kg of calcium, 2 kg of iron and 0·3 kg of phosphorus a year to each hectare.

In summary, therefore, we can see that the tropical rain forest ecosystem has a *quick and virtually leak-free mineral cycle* making it remarkably independent of deeper soil zone processes. This is shown in Fig. 5.7. Such a closed cycle of growth and decay means that the tropical rain forest is an ecosystem which man can upset and degrade quite easily.

Succession, climax and man

When a plant community reaches the kind of self-perpetuating state described above it is known as the *climax*. The tropical rain forest is therefore the climatic climax vegetation of the humid tropics. Climatic climax vegetation persists only when external conditions

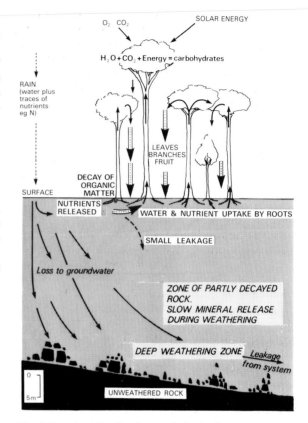

5.7   Mineral cycling in the tropical rain forest

remain unchanged—if climate alters or man clears the forest a number of changes occur.

With low human population densities the traditional method of farming in this zone was shifting agriculture. A patch of the forest was cleared and crops planted. As the recycling chain had been broken the crops quickly used up the available nutrient reserves of the soil. After gathering a few harvests man abandoned the land and moved on to clear a fresh patch of forest. The abandoned clearing soon becomes covered with a dense mass of weeds, grasses and shrubs. Tree seeds are also carried in by the wind and birds and a number of fast-growing tree species begin to colonise the area. As they grow they begin to dominate and shade out the lower vegetation. With the passage of time species more typical of the undisturbed forest begin their colonisation. The environment of the originally open clearing has been changed by the earlier colonists, becoming less extreme in its temperature and moisture variations, so that the dominant trees of the rain forest can succeed in growing. Eventually they shade out the earlier 'opportunist' species and the climax vegetation is re-established, after which there is no further change. Such an orderly progression is known as plant *succession*.

With increasing human population a second clearance may take place before the rain forest is re-established. In this case soil organic matter levels are further reduced, soil texture deteriorates and with the heavy rainfalls nutrients are washed out of the topsoil. When the shifting agriculturalists move on again after a few years of cropping, the soil is more impoverished than it was in the situation described above and the succession to rain forest climax may then be seriously impaired. At best a 'secondary forest' develops or at worst only grasses and bamboos can flourish where the changes in soil chemistry and texture have been more severe. Man in this situation may have induced soil changes to such an extent that progression to the rain forest climax is almost impossible. The succession is *arrested* by edaphic (soil-related) factors. Grasses and bamboo are also flammable and burning is the easiest means of clearance when cropping is required again. This means that rain forest seedlings, in their natural state not adapted to fire, are unable to colonise. Successful recolonisation also depends, of course, on the size of the area to be recolonised. Increased population means a greater proportion of the area cleared so that undisturbed rain forest patches may be some distance apart. Succession back to the climax may therefore be delayed because of the sheer size of the cleared areas.

Large areas of rain forest, particularly in the Indo-Malaysian formation, have been subjected to such 'slash and burn' farming for a considerable period of time. As populations have risen the time the land is left 'fallow' to return to forest has been reduced. Undisturbed primary rain forest is thus rare and 'secondary forest' and lalang (grass areas) now cover much of its former area. The Amazon Basin is currently the focus of clearance for agriculture and forestry and it seems likely that by the end of this century no substantial tract of 'natural' rain forest will remain.

Such disturbance involves various soil changes. Before examining tropical soils in a little more detail some other facets of tropical vegetation, climate and man will be discussed.

5.8   Secondary forest and agricultural clearings, Sri Lanka

The distribution of tropical vegetation: the roles of climate and man

Within the tropical world low temperatures are not a factor in determining the distribution of vegetation. It is, however, a different story when we consider rainfall. The rain forest environment has a high and well-distributed rainfall, but moving polewards the influence of the subtropical high-pressure cells is felt for part of the year. We thus have alternating seasons of moisture availability and aridity. Figure 5.9 shows the duration of the dry seasons within the tropics.

If Fig. 5.9 is compared with Fig. 5.4 the distribution of rain forest can be seen to correspond to those areas which have less than two and a half dry months. Where the dry season exceeds this we find a progressive shift in dominance from evergreen to deciduous trees (*i.e.* those adapted to periods of drought), a reduction in tree cover and stratification, and thirdly an increase in the proportion of grasses.

Where the dry season exceeds two and a half months a *semi-evergreen seasonal forest* is the climax vegetation. In comparison with the rain forest, trees are smaller, stratification is simpler (often with a deciduous canopy and an evergreen understorey) and there is a denser shrub and ground layer. The cerrado

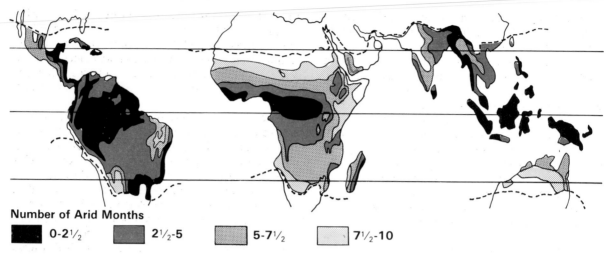

Number of Arid Months

■ 0-2½   ▨ 2½-5   ▨ 5-7½   ▨ 7½-10

5.9   Seasonality of moisture in the tropics

of Brazil and the monsoon forest of Indo-Malaysia are examples of this formation.

When the dry season is more protracted, *i.e.* about five months in duration, plant communities rather better adapted to drought conditions are found. The forest is dominated by deciduous trees, frequently with thick barks, small leaves and extensive root systems. The tree canopy is non-continuous and a single species, such as the acacia in Africa, may dominate. The ground is covered with tall upright perennial grasses. In this *deciduous seasonal forest*, a kind of grassy forest, during the dry season (*i.e.* before the arrival of the overhead sun) most of the trees and shrubs lose their leaves and the grasses die leaving a mass of combustible organic material on the ground.

With dry seasons of up to seven and a half months' duration, and a lower and more variable rainfall, *thorn woodland* is found. Here various low deciduous and evergreen trees and shrubs grow together with succulents separated by open grassy tracts. Adaptions to drought are more extensive, with trees having small leaves, scales and thorns to reduce transpiration.

Such woodland and grassland progressively gives way to *desert shrub* vegetation where the dry season exceeds seven and a half months. Individual plants are widely spaced, adapted either to conserve moisture (*i.e.* xerophytic) or by means of very deep root systems to tap deeper ground waters (*i.e.* phreatophytes).

So far no mention has been made of the term *savanna*, implying an open vegetation of grasses inter-spersed with trees. The co-dominance of grasses and trees in an open 'park-like' landscape is visually distinctive (Fig. 5.10). During the dry season such environments are almost desert-like in their severity but new foliage quickly follows the arrival of the summer rains. Grasses provide the dominant ground cover whilst scattered trees such as the picturesque baobab, acacias, euphorbias and palms dot the landscape.

The savanna biomes (*i.e.* a large area of vegetation of basic similarity) are floristically simple, consisting of only two vegetation layers: the discontinuous tree-shrub layer and a continuous herbaceous layer. Most of the woody plants are noticeably fire-resistant, being able to regenerate after fire, as can the tussocky grasses. The savannas have a large number and great variety of mammalian herbivores, such as antelopes, wildebeest, zebra and giraffe, preyed upon by a smaller number of carnivores. The summer rains impart a strong seasonal rhythm to plant, insect, animal and bird life.

The two *key questions about savannas* are, firstly, Why are they where they are? and secondly, Why are they like they are? In other words what explanations can be offered for their distribution and character. Climate, or more specifically the incidence of seasonal drought, is an underlying factor. However this changes gradually and in many cases the woodland savanna boundary is very sharp. Also semi-evergreen and deciduous forests on climatic grounds alone would also seem quite capable of occupying areas now dominated by various savannas, particularly in Africa. If the capability is there what has stopped this?

One answer may be *fire*. Man may have been firing such environments for almost half a million years as a hunter, for the last ten thousand years as a pastoralist and agriculturalist. Firing would destroy most tree seedlings and promote grass growth in the succeeding wet season. If you are a hunter grasslands provide a richer, more visible and more available herbivore food source. Expanding the supply of lusher young grasses would be a considerable advantage to herdsmen. Finally, clearing areas for cultivation by burning is less back-breaking than other methods and the ashes leave the surface with a supply of mineral nutrients.

Deciduous and semi-evergreen woodland with undergrowth burns quite easily in the dry season and continued firing would mean that it would be replaced by herbaceous and woody plants tolerant of fire. Thorny bushes and trees in such a situation would have particular advantages as they have natural protection against grazing herbivores. It can therefore be argued that savanna is a product of man's activities. He has arrested succession to perpetuate and extend an environment which he has found of particular use to him. Some support for these views comes from experiments which have protected plots from burning after which the successful growth of woody plants occurs.

At regional and local scales *other factors* besides man's activities may also be at work. In some locations flat plateau surfaces may be savanna-dominated whilst adjacent slopes are covered with forest. This could reflect the fact that fire runs ahead of the wind across flat areas. On the other hand it might reflect edaphic conditions. The flat plateau areas may be seasonally waterlogged in the wet season but dried out in the arid season. Sloping sites on the other hand favour evergreen forest because they shed excess moisture during the rainy season yet have access to migrating ground waters during the dry season.

The term *natural vegetation* has been deliberately avoided in this section for reasons which should now

5.10   African savanna

appear obvious. Even the concept of climatic climax vegetation is far from easy to apply as man may have interfered with natural succession by hunting, pastoral and agricultural activities, as has been discussed in the case of the savannas. This is the case even with the relatively low population densities and limited technology which have applied within the tropical world until relatively recently. This is one of the reasons why little attention has been placed here on the distributions of natural vegetation. A second factor is climatic change which will be discussed in the next chapter. Finally many vegetation maps for the tropical world have been based upon climatic characteristics rather than a detailed classification and mapping of actual vegetation. The emphasis in the discussion above has thus been placed on the interaction between man and nature. The references at the end of the chapter contain much fuller descriptions of both vegetation and its distribution.

SOILS

Soil composition and formation

So far in this chapter little specific attention has been paid to soils. Soils are natural bodies whose properties are 'due to the integrated effect of climate and living matter acting on parent material, as conditioned by relief, over periods of time'. This may not be the most elegant of definitions but it does concisely convey the complexity of soils which occupy such a central place in biospheric processes.

Soil has four *constituents*: mineral matter, organic matter, air and water. *Mineral matter* is derived from the weathering of the soil's parent material which may be bedrock or some transported material such as glacial or alluvial deposits. The various weathering processes either disintegrate this material into smaller fragments or decompose it (*i.e.* chemically alter it) so that new minerals are produced. Mineral material in the soil is classified according to its size. Particles less than 0·002 mm in diameter are clays, silts have particle sizes 0·002–0·05 mm, fine sands 0·05–0·25 mm and coarse sands 0·25–2·0 mm (US Grades). Particle size influences the *texture* of the soil which has important consequences. Sand, for example, tends to form an inert skeleton unable to hold moisture so that water drains freely and the soil is well aerated. The smallest particles, the clays, are crucial in controlling the physical and chemical nature of the soil.

Clay is composed of minute particles of silica, aluminium and variable amounts of oxygen and hydrogen left after various weathering reactions. The importance of the *clay fraction* in determining the physical and chemical properties of the soil stems from its properties. These include the ability to take up water, swelling when wet and shrinking when dry. Clay is also cohesive and this stickiness allows the formation of compound particles within the soil. It is also plastic, *i.e.* if it is deformed it retains its new shape. Finally, and of upmost importance, it is also chemically active.

This latter property stems from the fact that each tiny clay particle has a negative charge. To balance this electrically each particle is coated with a swarm of positive ions (cations). This loosely held swarm of cations includes a range of plant nutrients such as Ca, H, K, Na, Mg. (Ca is required for root tip and shoot growth, K in the production of carbohydrates and Mg is a basic constituent of chlorophyll, for example.) These nutrients move from the clay particle to the plant roots by the process of *cation exchange*. The plant root exchanges H^+ ions for the nutrient ions which it requires. The H^+ ions are often used in the further weathering of soil minerals leading to the release of more nutrient cations. This capacity for cation exchange is a fundamental process in the biosphere, second only in importance to photosynthesis.

The second component of soil is *organic matter* formed by the decomposition and digestion of plant tissues by various soil organisms. Decomposed organic matter, or humus, is colloidal with similar properties to the colloidal clays mentioned above. In fact it is about twice as effective as clay in terms of its cation exchange capacity and is therefore a very significant soil component.

The *texture* of soils is determined by the mixture of sizes of the primary soil particles (Fig. 5.11A). Clays and organic matter allow groups of sand and silt particles to join together into larger units—the *structure* of the soil—which affects its aeration and workability. These structures (plates, prisms, blocks and crumbs shown in Fig. 5.11B) are produced by pressure from wetting and root penetration together with the coagulating properties of the clay and organic matter.

Within the voids, between the primary particles themselves and between the units of the larger soil structures, we find the remaining constituents of the

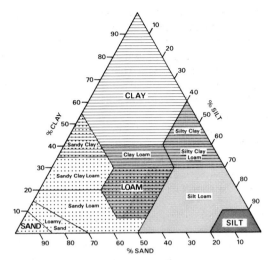

5.11A Soil textures

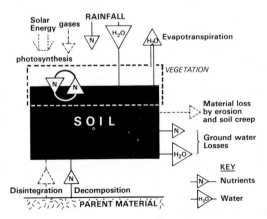

5.11B Soil structures

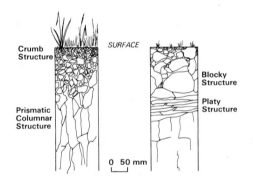

5.12 Inputs and outputs to the soil system

soil—air and water. *Soil air*, unlike the atmosphere, is usually saturated with water, richer in CO_2 and leaner in O_2. Some of the *soil water* is held as a thin film around the particles. This can be drawn upwards in connection with evaporation and transpiration and is known as capillary water. Gravitational water on the other hand is subject to the pull of gravity and moves downwards through the soil. As it moves the water can carry various chemical elements in solution and physically move clay and organic matter particles. The movement of water is therefore important in soil-forming processes. Whether soil water is acid, neutral or alkaline is also significant. When it contains a high proportion of H^+ ions it is acid and when it contains a high proportion of hydroxyl ions (OH^-) it is alkaline. This concentration is measured by the pH reaction of the soil, neutral soils (where H^+ ions and OH^- ions are in equal proportions) have a pH of 7, acid soils a pH of less than 7 and alkaline soils a pH of more than 7.

Soils are therefore complex mixtures of mineral particles, organic matter (dead and alive), water and nutrients. Figure 5.12 shows the *soil system* as a series of inputs and outputs within which a recycling of nutrients occurs.

From what has been discussed so far it should be obvious that parent material is not the sole determinant of soil character. Soil is in fact the product of five *pedogenic, or soil-forming, factors*—parent material, the type and amount of organic life, relief, climate and time.

Given the almost infinite number of possible combinations of the five pedogenic factors mentioned above it is hardly surprising that soils display such a variety of forms. The various systems of *soil classification* attempt to impose a collection of groupings on this diversity. One such approach involves the recognition of three soil orders: the zonal, intrazonal and azonal. The concept of *zonal soils* is derived from the work of the pioneer Russian pedologists who were impressed by the fact that they observed considerable parent material diversity, yet soil character and distribution seemed to parallel the vegetation and climatic zones. Zonal soils are thus viewed as those soils which occur over wide areas, on well-drained undulating land where the soil has been in place long enough for the climate and organisms to have actively expressed their full influence. *Intrazonal soils* were viewed as those which reflected some local condition, not involving climate or vegetation, such as poor drainage. Finally, *azonal soils* were young soils lacking a well-developed soil profile. These terms will be used in the sections which follow but it is important to note that all classifications are merely systems of convenient groupings and for some purposes a classification reflecting capability for agriculture may be required.

Tropical soils

As tropical vegetation has already been discussed we will begin our examination of soil in the context of the tropical world. *Latosol* is the term applied to the broad group of soils found in the rain forest and savanna zones. Only a small proportion of this area is underlain with basic rocks, those rich in ferromagnesian minerals and low in silica. Although accounting for probably only 2% of this environment the soils developed on this basic rock are capable of sustaining continuous cropping and are more important than their area would initially suggest. Under the high temperature and moisture conditions rock and mineral weathering proceeds rapidly. Although weathered material is dissolved and washed down through the soil (the *leaching* process) weatherable material is available in the soil and minerals are released continually. A group of soils termed *basisols* are associated with such basic rock areas. One example is the *ferrisol*, red in colour, with an adequate organic matter content and a well-developed blocky structure. However, such soil is restricted in occurrence and far more representative of the latosols are the leached ferrallitic, ferruginous and weathered ferrallitic soils.

Where rainfall exceeds 1,500 mm in the tropical rainy climate, under rain forest and the semi-evergreen seasonal forest, the *leached ferrallitic soil* is the zonal soil type. These are highly weathered, lacking weatherable minerals within the frequently very deep soil profile. They are strongly acid (pH 5·5) containing up to 3% organic matter under the rain-forest vegetation, and range in colour from red to yellow. With the high temperatures and more than adequate moisture minerals quickly decompose. The silica (SiO_2) becomes mobile and is leached out of the soil. This leaves the sesquioxides (Fe_2O_3 and Al_2O_3) in the soil which gives it its characteristic colouration. The clay fraction is dominated by kaolinite, the clay mineral with the lowest cation exchange capacity. Their almost closed leak-free nutrient cycle was outlined on page

119 and in Fig. 5.7. It is not surprising, therefore, that sustained cropping seriously depletes the nutrient reserves of this soil.

The *ferruginous soils* are found outside of the rain-forest areas and represent the zonal soils of the wet and dry tropical climate, *i.e.* those zones covered with savannas and the deciduous seasonal forests. Such soils are moderately to highly leached, having pHs between 5·5 and 6·5 (moderately acid). Typically shallower than the leached ferrallitic soils and ranging in colour from red to reddish brown they can sustain continuous cropping. On plateau surfaces within the savanna zone where there has been deep weathering of the regolith, *weathered ferrallitic soils* are found. These are deep, only moderately leached and with pHs of less than 5·5. Paler in colour than the ferruginous soil they have less organic matter, too (*i.e.* 1% compared to up to 2% for the ferruginous soils). They have a lower agricultural capability and are unsuitable for continuous cropping.

In areas of poor drainage, such as valley floors, alternating oxidising and reducing conditions are present according to the variations in soil-water levels. In such situations the intrazonal *gleys*, mottled dark-grey clays, are found. Such sites are frequently utilised for pasture and padi.

Within almost any latosol, layers of *laterite* may be found, occurring as a massive layer up to 10 m in thickness or in nodular concretions. This material (which is not a soil) is iron in its ferric form. Being resistant it can be a significant limiting horizon in the soil, preventing root penetration and influencing drainage.

Models of tropical ecosystems

Nutrient cycling has already been mentioned several times and this movement of minerals within the rainforest and savanna ecosystems can be simplified as a series of stores and transfers. Such a model consists of the biomass store (the plants and organisms) which, as tissues die, contributes material to the litter store. As the decomposer chain operates this litter store releases nutrients to the soil store, where they are available for plant uptake. The losses from the system are compensated for by inputs from rainfall and from rock decay. Without such recycling of nutrients and the replacement of losses, mineral starvation would cause the ecosystem to die.

The rain-forest model consists of a large biomass

store and smaller litter and soil stores. This is shown in A of Fig. 5.13. With no drought or frost, transfer between the stores is rapid and continuous. When 'slash and burn' farming takes place biomass is removed by felling and burning and nutrients are transferred to the litter and soil. Nutrients (and energy) are also diverted by harvesting and the modified ecosystem is shown in B of Fig. 5.13. In the case of rice padi the flooded fields retard litter and decomposition, runoff loss is reduced and nutrients are retained in the litter and soil stores (C of Fig. 5.13). Man in both cases alters the system to enrich the soil although this is only temporary in the case of B of Fig. 5.13. The savanna system, with its smaller biomass, proportionately larger litter and soil stores and reduced leaching loss is shown for comparison in D of Fig. 5.13.

MID-LATITUDE CONTINENTAL INTERIORS

In this section inter-relationships between the living and non-living environments will be examined using

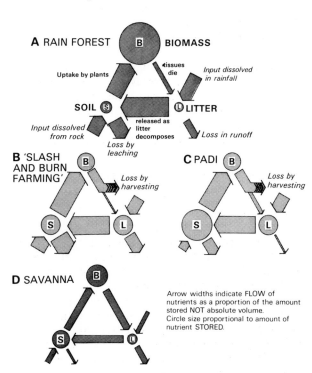

5.13   Ecosystem nutrient cycles

the Province of Saskatchewan as an example. Some aspects of its water resources were introduced in Chapter 4. It is returned to here as it provides examples of two major world biomes, the coniferous forest and the temperate grassland.

The coniferous forest zone

The northern part of Saskatchewan consists of an exposed part of the Canadian Shield whose Pre-Cambrian rocks are mantled towards the south by younger sedimentaries. Both areas have been glaciated and it is therefore the legacy of the Pleistocene which determines the detailed terrain on which plant and soil development takes place. Approximately 20% of the area consists of lakes, either eroded rock basins or hollows produced by irregular deposition of glacial material. Dryland sites range from exposed crystalline and sedimentary rock to an extensive and diverse collection of glacial gravels, sands and clays. Although there are no pronounced relief features local undulations in the order of tens of metres can be quite pronounced.

This northern zone of Saskatchewan experiences a subarctic climate. July means reach 18°C but there is a longer winter with January means around −20°C and winter snowfalls of 150 cm. Coniferous trees, adapted to such an environment form the climax vegetation. In contrast to the rain forest it is floristically simple, with only a few tree species present in any one area. Its primary productivity is also lower compared to the rain forest.

Different tree species have varying tolerances and adaptions to both climatic and soil factors. This produces the zonation of forest shown in Fig. 5.18 with black spruce (*Picea mariana*) dominant in the north. The more demanding white spruce (*Picea glauca*) occurs only on warmer sites in this zone such as south-facing slopes. Towards the south it becomes more prominent on deeper soils and well-drained sites whilst the black spruce is restricted to the wetlands. Where disturbance has occurred, through fire or logging, extensive stands of jack pine (*Pinus banksiana*), an aggressive pioneer tree, may occur.

Figure 5.14 models the nutrient circulation for the coniferous forest ecosystem. Compared with Fig. 5.13A its distinctive feature is the large litter store and small soil store. How is this produced and what consequences has it got for soil formation?

Coniferous trees make only small demands on the

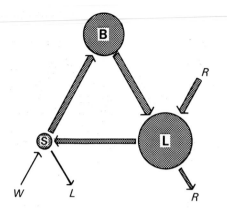

5.14   Coniferous forest nutrient cycling

supply of bases from the soil so the litter supply is low in calcium and somewhat acid. In the shade beneath the trees, frozen all winter, the decomposers have rather a hard time! Litter therefore accumulates on the surface, being transformed slowly into an acid (or mor) humus. The spring thaw allows water to percolate through this litter where it becomes charged with a large supply of organic compounds. Passing downwards these compounds detach mineral ions from the

upper layers of the soil. This is known as *chelation*, a more dramatic process than leaching, since it removes plant nutrients, Fe and Al. The upper layers of the soil are therefore eluviated, or have material washed out of them, leaving a bleached colourless layer of SiO_2. In the lower part of the soil profile humus, Fe and Al are washed in (illuviation) and redeposited. These processes give a soil profile with a series of very distinct layers or horizons, the podzol.

The process of *podzolisation* and its distinct horizons are shown in Fig. 5.15. It is a widespread soil-forming process and reaches its best expression where water can move easily down through the soil (*i.e.* where precipitation exceeds evaporation and where soil textures are free-draining and where the surface is mantled with an acid, base-lacking litter).

Northern Saskatchewan is a complex area of lakes, bogs, patches of forest and rock outcrops. The lakes in the long term are only temporary features and a process of bog succession is occurring. The concept of succession has already been mentioned in the context of secondary forest within the rain forest. Figure 5.16 is an example of *autogenic succession* where the plants themselves are progressively changing their environment so that more demanding species eventually become dominant.

With relatively stagnant and acid surface waters peat mosses may in time support open stands of small

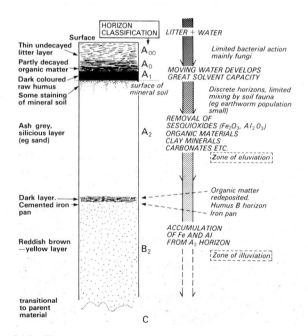

5.15   The development of the podzol soil profile

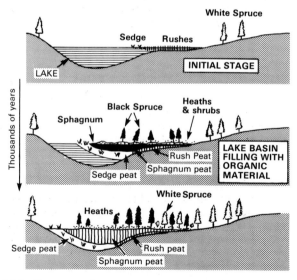

5.16   Bog succession

5.17 A string bog (B) bordered by spruce woodland

black spruce and tamarack. Such 'treed bogs' are known as *muskeg*. In situations where there are very gentle slopes together with a lateral movement of water, less acid peat-forming mosses and sedges colonise the area. This leads to the development of raised organic ridges and intervening wet troughs, the so-called *string bog* (Fig. 5.17). Where the raised ridges are relatively dry an open cover of tamarack, willow and birch may colonise. A particularly large area of wetland is shown in Fig. 5.18A where the Saskatchewan River forms an inland 'delta'. It forms a complex area of rivers, levees, cutoffs and flood plains with a mosaic of mixed and coniferous woodland on the drier sites between the muskeg and bogs.

Prairie grasslands: soils and organisms

The southern part of Saskatchewan experiences a climate typical of its continental interior location. Mild summers and cold winters reflect its latitude and because of its interior position, removed from maritime influences, it has a wide temperature range (*i.e.* January −15°C and July 20°C) and a low precipitation, 70% of which occurs in summer. Precipitation varies from year to year and potential evapotranspiration exceeds precipitation. This southern zone of the province is thus too dry for trees, apart from favoured sites in valleys and around depressions. Grassland formed the climax vegetation over most of the zone. In the extreme south-west the Cypress Hills, rising 600 m above the general level, and being moister and cooler (Fig. 5.18A) originally sustained a forest cover.

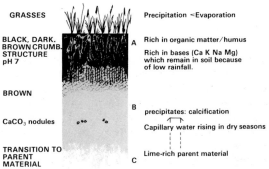

5.19 Prairie grassland nutrient cycling and soil profile

5.20 Saskatchewan, mid-grass prairie

Soil development and nutrient cycling under grasses contrasts markedly with the coniferous forest situation described earlier. At the end of winter the soil is damp, progressively drying during the growing season. As grasses die they add organic matter to the surface and their mat of fine hair-like roots are also incorporated as part of the soil's humus. Grasses bring up bases from within the soil and these are recycled by this litter and root decay. As annual evapotranspiration exceeds precipitation there is no leaching or removal of bases from the soil profile. The bases in fact concentrate as precipitates in the lower part of the B horizon, in this area at depths of 75–120 cm. The zone of maximum biological activity, the black-dark brown A horizon, is typically 25 cm deep. Rich in finely disseminated humus its organic matter content

ranges from 8 to 30% and it usually displays an excellent crumb structure and a neutral pH 7 reaction. This is the *black earth* or *chernozem* soil whose profile and nutrient cycling model is shown in Fig. 5.19. It stands in strong contrast to the podzolisation regime which occurs in the cooler and damper environments to the north.

As can be seen from Figs. 5.18A and 5.18B the location of this black soil corresponds to the distribution of the fescue grasses and aspen grove. Moving south into areas with less effective moisture there is a transition through various types of 'mid-grass' prairie, for example the spear grass blue grama wheat grass prairie. This is a mixture of grasses whose annual growth cycle is suited to the climate, beginning to grow in April and with their development virtually complete

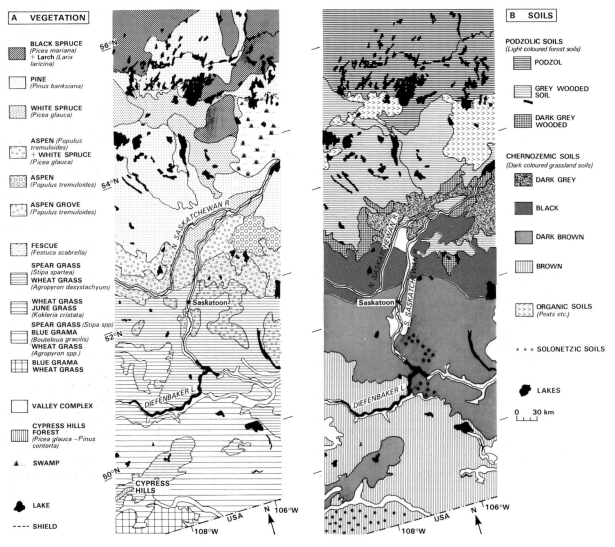

5.18   Vegetation and soils in Saskatchewan

by the period of moisture deficiency which begins in July. In the driest areas of the south-west of the province blue grama wheat grass is dominant and grass growth is stunted in most years becoming somewhat taller only in moist years.

If Figs. 5.18A and 5.18B are compared the vegetation and soil belts can be seen to be in broad correspondence. The *dark brown* (or chestnut) *soils* are similar to the chernozem but contain less humus and their lime accumulation zone reaches to within 60 cm of the surface. The *brown soils* are even lighter toned, reflecting their lower humus content from the less luxuriant grasses, and with higher temperatures and evaporation rates the lime accumulation zone may reach to within 35 cm of the surface. The black–dark-brown–brown sequence of soil belts mirrors increasing aridity and reduced biological activity. They are developed across a range of parent materials (clays, loams, silts), reflecting a complex of glacial deposits and are thus an example of *zonal soil* distribution.

In some locations the mantle of glacial deposits is thin and underlying shales are exposed. In these situations salinisation may occur where salt solutions (*e.g.* $NaCO_3$) are drawn upwards to the surface where the water evaporates leaving alkaline salts in the soil. We therefore have the development of *solonetz* soils, with tough, hard, prismatically structured B horizons and patches of salt on the surface itself. Such soils reflect the operation of parent material influence and are examples of *intrazonal* soils.

Dynamics of the prairie ecosystem

Endless level fields of grain and grazing beef cattle encapsulate many people's image of the prairies of North America. The misconceptions about farming, let alone terrain, that such images convey are beyond our scope here but these mid-latitude temperate grasslands are nonetheless the location of significant livestock and grain production. It is important to note, however, that their primary productivity is only a sixth of the savanna's and with the usual sequence of energy loss through the ecosystem only about 0·004% of the incoming solar energy ends up as meat 'on the hoof'.

When the savanna was described, the role of man in influencing the character and distribution of these tropical grasslands was stressed. As will be seen in the paragraphs which follow it has been a similar story in the case of the prairies. Searching a little deeper, however, involves assessing the roles of organisms and climate.

The interaction of all the environmental factors can frequently best be seen at the boundary of two vegetation types, *i.e.* at the junction of grassland and trees. The aspen (*Populus tremuloides*) is the dominant tree at the forest–grassland boundary in Saskatchewan. Close to the transition it often forms pure stands of fifteen-metre-high, closely spaced trees. To the north increasing cold means that it becomes mixed with various coniferous species. Towards the south, with increasing aridity, the aspen becomes smaller (three metres in height), the pure stands fragment into groups of trees or groves and finally it is found only in moister depressions and around sloughs. The *aspen grove* itself (see Fig. 5.21) consists of groups of trees, often with a dense undergrowth of shrubs and herbs, separated by stretches of fescue grassland.

The distribution and character of the aspen grove reflects the interaction of a range of factors. The boundary between areas moist enough and too dry for tree growth is, of course, a zone rather than an abrupt line. Within such a zone a mixture of groves and grassland may have been in some kind of equilibrium with available moisture supply. The groves would have concentrated drifting snow, thus increasing their moisture supply and chances of growth and survival during the following summer. A mixture of trees and grasses may therefore have been stable in an area not moist enough overall to support a continuous tree cover. With cycles of dry and moist years, the situation would have become a little more complex, with grove expansion during wetter periods and contraction during drier periods.

Fire too may have been a significant factor in eliminating trees and maintaining grasses. However, it is possible to ask if it is that simple. Aspen grove spreads in two ways, by seeding and also by regeneration from its underground root system. The latter method may have been quick enough to ensure tree survival before a mat of grass recovered a burnt area.

The larger herbivores also played their part. There were probably upwards of 10,000,000 bison in the Canadian section of the Great Plains. Grazing in large herds these would have reduced the fescue and spear grasses, leaving the shorter and mid-grasses (wheat grass, grama and June grass). One aspect of their behaviour in particular may have aided the spread of aspen. In spring, after the snowmelt, the increasing warmth and surface moisture saw a tremendous explosion in the numbers of biting insects. The bison, losing its winter pelage at this season, was particularly susceptible to mosquito harassment and to seek some relief they wallowed in the soil. Being large and heavy beasts this wallowing created shallow depressions about 6 by 12 metres. These coalesced to form dusty basins stretching for kilometres. Such bare soil sites became colonised by shrubs such as the snowberry. Aspen seeds carried by birds would also colonise such sites and stand a good chance of survival with no competition from grass and with winter snowfalls accumulating in the shrubs. This sequence shows that grazing animals do not merely eat and destroy vegetation (like the elk and deer in the case of the aspen groves themselves). They can also create opportunities for future plant growth.

So far we have seen, therefore, that aspen grove was in a kind of *dynamic equilibrium with fire, climate and animals*. The question now remaining of course is what was the *role of man*?

Indians were part of this ecosystem, too, hunting the bison herds in summer, for example, and using fires and decoys to drive them to the 'pounds' over steep valley-side cliffs. Following the arrival of the European in North America—or more specifically his metals and horses—the Indian could hunt selectively for the first time. Killing female bison for their tender flesh marked the first tentative steps towards the animal's final extinction. The classic Plains Indian culture familiar to us all from film and television was relatively short-lived. The first major assault on the ecosystem, however, awaited the spread of the fur trade during the late eighteenth century when bison were slaughtered to provide dried meat or pemmican, a source of durable and portable protein for the trappers and traders. The reduction in the herds continued during the nineteenth century, reaching its climax after 1871 when tanning techniques allowed the previously soft bison hide to be competitive with cowhide for the first time. By about 1883 the majestic herds on the Canadian Prairies had followed their American cousins to extinction. Each 'turn of the screw' or increase in hunting pressure had made the beasts more susceptible to such natural factors as wolf and bear predation, disease, hard winters and fires.

5.21 Aspen grove, northeast of Saskatoon

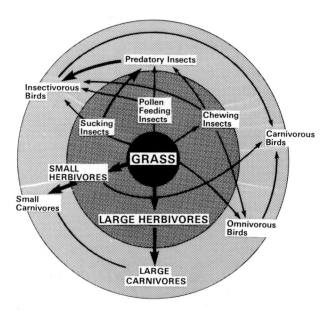

5.22 A simplified grassland food web

The original grassland ecosystem is shown in terms of its simplified food web in Fig. 5.22. Losing its dominant herbivore would have naturally had effects on both vegetation and the other animals.

Before the expansion of the railway network across the Canadian Prairies the next few decades saw the spread of *ranching* in many areas. Although the rancher used resources extensively he was nonetheless attempting to manage the ecosystem. His cattle replaced the bison as the dominant herbivore and bearing in mind the kind of relationships shown in Fig. 5.22, the rancher tried to firstly reduce competition for grass and secondly reduce the attention of predators. Climatic fluctuations, particularly the winter of 1906 brought a realisation of the need for supplementary feed (hay) and the beginnings of irrigation.

Arriving later the *farmer* changed the ecosystem much more fundamentally. With his perception of the prairies conditioned by values and farming practices appropriate to humid eastern Canada or Europe, the process of adjustment to the realities of the prairie world were a little hesitant. This process was not aided by an initially moist period which was followed by drought. The removal of the bison had also led to the growth of taller grasses which overemphasised the moisture of the prairies. In general the farmer viewed the natural landscape as one to be tamed. The prairie sod was broken by the plough and its nutrient cycle broken. Such a break would ultimately demand its replacement by fertilizer applications, fallowing and crop adaptions. Intolerant of competitors—'weeds'—his grain crops replaced the mixtures of natural grasses. This monoculture was exposed to the risk of insect infestations, for example by grasshoppers. Over the last sixty years prairie farmers have developed more respect for their place in nature's system. Although the simplified farming ecosystem is less robust than the natural one it replaced, techniques of dry farming and a more diversified crop and livestock base have better equipped it to face the fluctuations of the prairie environment as well as changing economic climates.

If you have read this story surrounded by the asphalt and concrete surfaces of the city it may seem somewhat irrelevant. The details of aspen grove, bison, ranchers and farmers are certainly space and time specific yet the general points which have been made are not—*nature is seldom simple and man has to be seen as part of the ecosystem.*

THE BRITISH ISLES

The soils of North Somerset

This small area has been focused on earlier in the book. It has been chosen as the final regional example because it shows many general points about the nature and distribution of British soils.

Although England's deciduous forest climax vegetation has long been cleared, the zonal soil group associated with it, the *brown earths*, still remain. A typical brown earth soil profile and the processes which have produced it are shown in Fig. 5.23.

Figure 5.23 can usefully be compared with the soil-forming regimes and mineral-cycling models shown earlier in Figs. 5.13, 5.14 and 5.19. One contrast with

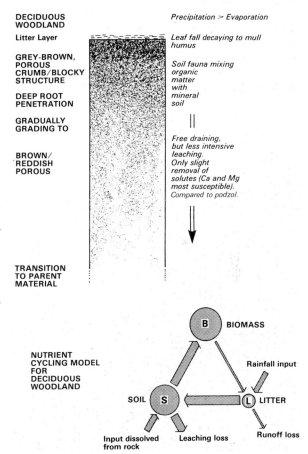

5.23   Brown earth soil

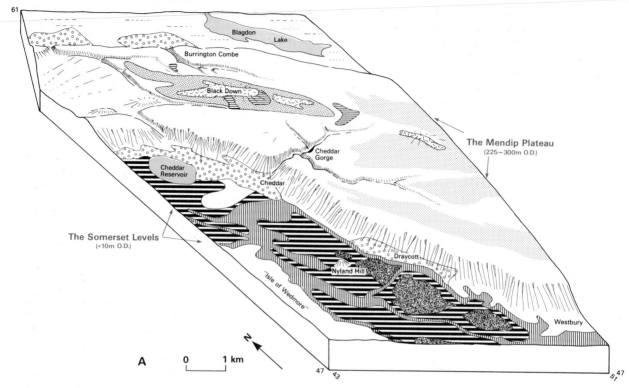

TOPOGRAPHY	PARENT MATERIAL	FREE DRAINING SOILS	IMPERFECT DRAINAGE	POOR – VERY POOR SOIL PROFILE DRAINAGE
		BROWN EARTHS		
Mendips, plateau and slopes	Silty Drift over Carboniferous Limestone	Nordrach		
				PEATY GLEYED PODZOL
	Old Red Sandstone	Maesbury		Ashen
	Old Red Sandstone, solifluction (head) deposits			**SURFACE WATER GLEY**
				Thrupe
Lowlands	Gravel solifluction (head) deposits over Keuper Marl.	Langford		
			GROUND WATER GLEY	
Somerset Levels	Riverine clay from Keuper rocks.		Compton	
	Estuarine clays and alluvium.		Allerton	
		ORGANIC		
	Fen peat over estuarine clay.	Godney		

SOIL GROUPS (Including podzol and gley sub groups)

5.24   The soils of north Somerset
   A Topography and the distribution of eight soil series
   B Topography, parent material and drainage class of
      the soil series mapped above

podzolisation, for example, is that although both soils reflect free-drainage conditions, the leaching processes in the brown earths are subdued. Organic matter from leaf fall is mixed into the upper horizon by the high population of earthworms, thus avoiding the build-up of an acid litter layer on the surface.

The Soil Survey of Great Britain until 1974 classified soils into six *groups*, brown earths, gleys, podzols, calcareous soils, organic soils and undifferentiated alluvium. Within these groups the mapping unit used was the *soil series* within which there were common profile characteristics. The soil series were named after the location where they were first described or best developed. Figure 5.24 shows the distribution of eight soil series belonging to four groups. Within the area shown on the diagram the Soil Survey recognised forty-six series so only a selection of these are shown.

Figure 5.24 portrays a 130 km² area of north Somerset. Topographically it consists of three major divisions—the Somerset Levels, their fringing lowlands and 'islands' and thirdly the Mendip plateau. The Somerset Levels, the extensive flat areas below 10 m OD, are dominated by two soil groups, the various belts of groundwater gleys and patches of organic soils. *Gleys* are intrazonal soils occurring in water-logged sites—in other words they are hydromorphic soils. Water filling the pore spaces within the soil quickly becomes de-oxygenated and in this anaerobic environment bacterial decay is slowed down. Under these conditions iron within the soil becomes reduced and in this ferrous state it imparts the bluish/greenish/grey tinge characteristic of the gley soils. Root passages and cracks in the soil occasionally allow the penetration of fresh oxygenated water or, of course, air itself during the drier seasons. In these situations the iron oxidises and gives red/yellow patches of colour (mottling). Above the level of waterlogging, oxidation gives the usual brownish and reddish colours to the soil although these are muted by darker-coloured humose layers near the surface. The *groundwater gleys* are soils in which the gleying occurs because of the pattern of regional drainage which concentrates ground water in low-lying or receiving sites.

Figure 5.24 shows the distribution of two soil series belonging to this groundwater gley group. The Compton and Allerton series both show similar gleyed profiles but they have textural variations reflecting the influences of their different parent materials. The Compton series, texturally a loamy clay with a neutral-acid reaction, occurs *on parent material* of Keuper age which has been moved and redeposited by stream action. The Allerton on the other hand has developed on estuarine clays which were deposited in association with marine transgressions when sea-levels in the Bristol Channel, a few kilometres to the west, rose and fluctuated after the Pleistocene period.

As a process reflecting fluctuating water levels within the soil, gleying is common in river valleys, along flood plains and at the foot of slopes. Such 'meadow soils', as they are sometimes known, are a common component in the British soil scene.

Organic soils, as can be seen from the distribution of the Godney series in Fig. 5.24, occur as patches away from the stream lines which were reflected in the distribution of the Compton series. In such poorly drained situations a water-tolerant vegetation dominated by reed grass (phragmites) developed on top of the estuarine clays. This gradually filled the basins in a process analogous to that shown earlier in Fig. 5.16. These waterlogged sites therefore saw the accumulation of peats. With continued accumulation of plant material the peat-bog surface gradually became raised into a gentle dome. Its surface became drier and it progressively became invaded by different types of vegetation culminating in a dense low woodland (carr) interspersed with bog areas.

Over the last few centuries such 'natural' vegetation has disappeared. The annual flooding of the Levels has been reduced by the construction of a network of drainage ditches (rhynes) which connect with artificially straightened main drainage ways which in turn are pumped into the Bristol Channel. This lowering and stabilising of the water table has meant that the carr and marsh have been replaced by permanent pasture. However, during the last decade economic pressures have caused a reappraisal of the agricultural potential of the Levels. The peat itself had been dug and dried for generations on a small scale for local fuel. In some parts of the Levels it is now removed for packaging as horticultural peat, to be applied as organic matter to gardens over the breadth of southern England. It has also been realised that if the water table could be lowered still further these organic soils could support profitable arable farming. At the time of writing this latter development is restricted in area. It is important to note that with a precipitation twice that of the fens of eastern England the arable crop combinations would be a little different. The economic pressures of modern farming operating within the rules of the EEC agricultural policy may, however, mean that the landscape of permanent pasture and straight rhynes bordered by pollarded willows so characteristic of the Somerset Levels, may be unrecognisable in a few decades.

As has already been stressed *brown earths* form the zonal soil group on free-draining sites. Only three soil series belonging to this group have been plotted on Fig. 5.24. The Langford series occurs on lowland sites fringing the Mendips. Its parent material consists of 'head' (fans or spreads of detritus washed, slumped or flowed down from the Mendip Hills during the Pleistocene) overlying red Keuper marls. These soils are neutral-acid, brown to grey-brown in colour and loamy in texture although varying amounts of sands and stones may be present in the profile.

On Mendip itself are extensive spreads of the Nordrach soil series. This is a freely drained, acid-neutral, dark red-brown silt loam in its upper horizons which passes into a reddish brown clay in the lower part of the profile. Since the bedrock is Carboniferous Limestone and the Nordrach series is found on the top of the plateau it is reasonable to ask where the silt has come from. There are no rivers to carry it there. Also the silt-sized material produced by the weathering of of the limestone bedrock does not display the same mineral assemblages. Since the grain sizes of the material involved is consistent with wind transportation it may in fact represent a *loess* deposit. Such wind-transported material would have been moved and deposited during the Pleistocene period when the Mendip plateau would have enjoyed a tundra-like periglacial environment (discussed in Chapter 6).

The third brown earth plotted is the Maesbury series. If you look at Fig. 5.24 and recall the geology of this part of Mendip (pp. 54–8) you will realise that it is developed on the quartzitic Old Red Sandstone. This has weathered to give a porous, very acid, brown, sandy loam with plenty of stones in the soil.

Fringing the Old Red Sandstone inlier of Black Down there are also discrete patches of the Thrupe soil series. As the key of Fig. 5.24 shows, this is another gley soil, but in contrast to the groundwater gleys mentioned earlier this is a *surface-water gley*. The excessive water is on the surface and gleying occurs in the upper horizons above a kind of 'perched' water table. Beneath this there is some gleying and the lower horizons are brown-coloured. The Thrupe series has developed on Old Red Sandstone head or colluvium washed down from the higher slopes.

The final soil type whose distribution is shown in Fig. 5.24 is the Ashen soil series, a *peaty gleyed podzol*, found on the upper summit areas of Blackdown. Peaty gleyed podzols are common soils in upland Britain where rainfall totals are high. Peaty gleyed podzols have poor surface drainage and quite pronounced texture differences within their soils. As can be seen from Fig. 5.25 there is a peaty A horizon beneath a raw humus layer. Beneath this acid leaching occurs resulting in a bleached A2 horizon. Above the iron pan waterlogging and gleying occur.

British soils: factors in soil genesis

The detail in the section above was intended to introduce some characteristic British soils. The size of area examined has been much smaller than in the earlier parts of the chapter. The brown earths, gley, podzol and organic soil groups reflect different combinations of soil-forming processes. As the individual soil series were differentiated on the basis of textural differences

in their upper horizons, it is not surprising that such a classification system results in mapped soil series whose distribution mirrors the *parent materials*. It should be remembered, however, that the bedrock and transported materials constituting the parent materials are only one of the soil-forming factors.

In many situations regular sequences of soil changes downslope are found so that identical soils are found in similar topographic positions. This chain-like pattern is known as a *catena* and an example is shown for the Mendip area in Fig. 5.26. The soils are genetically different (*i.e.* different in their origins) and vary in their profiles but they do recur in regular sequences wherever the same topographic conditions are encountered. The catena thus illustrates the importance of *slope and drainage* factors in soil genesis.

The role of *plants* in soil development is complex. Soil characteristics like depth, structure, texture, moisture, nutrient supply and acidity influence the type of plant community which develops. Plants in turn remove nutrients, add organic matter and increase aeration. Mention has already been made of this interaction and the nutrient cycling models used in the chapter have been designed to emphasise this general point.

One specific example in the context of the situation shown in Fig. 5.26 is the relationship between the heath vegetation and peaty gleyed podzol on the upper, poorly drained sites. The water in such loca-

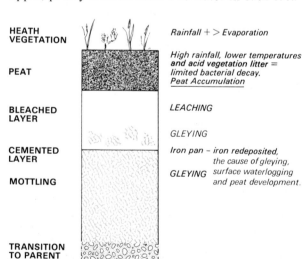

5.25 A peaty gleyed podzol

tions contains high concentrations of iron, aluminium and manganese. The hard and glossy leaves of rushes (*Juncus*) and cross-leaved heather (*Erica tetralix*) with their thick cuticle and small stomata are xerophytic features reducing transpiration losses. With reduced water loss toxic concentrations of these minerals can be avoided: if they freely transpired mineral build-up within their tissues would kill the plant. In the case of heather (*Calluna vulgaris*), another plant restricted to acid sites (*i.e.* a calcifuge), it appears to be able to produce organic compounds which include the iron and aluminium. By 'locking them up' in this way the heather is able to grow in acid conditions. The other side of the coin is that when the heather dies they are released again, encouraging chelation and thus podzolisation. Conifers have the same kind of effect and under Forestry Commission plantations soil acidity has increased by this process of cation exchange. There are, of course, many other examples of vegetation–soil relationships and these can be found in the further readings listed at the end of the chapter.

Discussion so far has related to the influence of parent material, topography and vegetation. There is, of course, the important dimension of *climatic influences* on soil variations. This was the focus of some of the earlier discussion on tropical and prairie soils. In the British context the climatic influence is primarily expressed through the role of *water* and its effects are summarised in Fig. 5.27.

A final factor is *time*—just how long does it take for soils to reach some kind of equilibrium with their environment? This period is probably in the order of

centuries rather than decades but since soil science is a new field of study this remains an open question. To complicate answers, over the long term we have the question of climatic and vegetation changes and in the shorter term the effects of man.

MAN, SOIL AND THE BIOSPHERE

Man's influence on soil has grown as his numbers and technology have increased. In broad terms man's impact on soils can be considered under two headings. Firstly man can modify and manage plant succession through grazing and other low-intensity uses, a point which has been made earlier in the chapter. In the British context upland heaths and moors might appear as 'natural' vegetation to an uninitiated urban eye. This is far from the truth. In the case of grouse moors, for example, fire is an integral part of their management. Burning patches of about a hectare, with six such patches to the square kilometre, produces the highest grouse populations, as the birds require unburnt heather areas for nesting yet find far more food in the regrowing patches. In the case of the chalk downlands of England sustained sheep grazing has produced a dense fine turf of sheeps fescue (*Festuca ovina*). Such interference with succession produces various kinds of *plagioclimax* vegetations which in

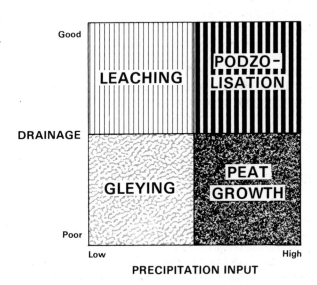

THE CATENA

| PEATY GLEYED PODZOLS | BROWN EARTHS | GLEYS |

Ashen series

Maesbury series

Ellick series (not mapped on figure 5.24)

Thrupe series

Flattish summit surfaces. Poor drainage. Heath vegetation. Peat accumulation and gleying.

Slope better drained.

Impermeable bedrock, downslope water movement through soils leading to waterlogging and gleying.

5.26 A soil catena: Blackdown, central Mendip

5.27 Water balance and British soil processes

turn have consequences for nutrient cycling and soil development.

Secondly, of course, man may remove the natural plant cover and replace it with crops. Centuries of ploughing and draining together with more recent compaction by farm equipment produce changes in soil structure and the arrangement of horizons. Replacing a climax forest with arable cultivation breaks the natural cycling of nutrients shown in Fig. 5.23. The loss of leaf fall affects perhaps the most vital component of a 'good' soil—its structure. Liming to reduce acidity and applying fertilizer to replace nutrients are therefore only part of the story of successful management of brown earths. Incorporating ley grass in the rotation attempts to replace the organic matter which enabled structural aggregates to form and promote drainage and aeration within the profile.

Human food supply is the final product of chains which begin with primary inputs of solar energy, land and water. To these natural inputs man adds plant nutrients, pesticides, seed (derived from thousands of years of trial and error and a much shorter period of plant breeding), his labour, fossil fuel energy and capital. These inputs and four pathways from the actual plant growth are shown in Fig. 5.28. The left-hand pathway represents a 'vegetarian' chain and the other three, various livestock chains. The figures at the bottom are the amounts of solar energy which eventually end up as human food and they reflect two points which have been made earlier in the chapter. Firstly, that all ecosystems have energy losses at each stage, reflecting the efficiency of photosynthesis and the requirements of organic growth. Second is the difference between the four chains in respect of the amount of solar energy converted into human food. The latter is obviously relevant to the population resources debate.

A simplistic reaction is to focus on 'waste' in the livestock chains and to propose a shift to a vegetarian diet. This ignores two facts, the nutritional qualities of animal protein and secondly the role of livestock in sustaining farm systems. When entering the debate it is important to consider firstly the meaning and accuracy of definitions and statistics and secondly the effects of any changes. Food, for example, can be defined in terms of calories and vitamins, yet most British people don't eat horse meat and might be quite upset at the prospect of eating dogs. In other words food is culturally referenced, there may be religious

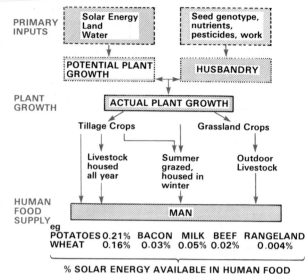

5.28   Agricultural food chains

beliefs or socially transmitted preferences which determine what man will eat compared with what he could eat.

Our attention is often focused only on the food production side. An overconcern with fields tends to blind us to the storage, distribution and preparation chain where malfunctioning of the social and economic system may produce losses and scarcity which are human in origin and not naturally induced.

Reliance on a limited gene stock for seeds in 'engineering' higher yielding strains may have side-effects. Plant disease, infestation or climatic failure can now affect much larger areas than in the past when more varieties were sown. The 'model' of energy-intensive and mechanised European and North American agriculture, besides being ecologically suspect, may also produce economic and social side-effects when applied in the developing world. It could swell a country's oil import bill and decrease its rural work opportunities.

In the space of a few paragraphs it has been possible to merely scratch the surface of this problem. Geographers have sometimes been guilty of a limited view on this global question. Our concern with the distribution of soils, climates and people ought not to blind us to the interdisciplinary nature of any meaningful approach to this vital issue.

In conclusion and to sum up one of the themes of this chapter it seems appropriate to remind ourselves 'there's no such thing as a free lunch'. Nature's networks are far from simple, whether they occur within a few hectares of the tropical rain forest or at the global scale. Man now has some understanding of the energy flows and material cycles of the earth but rarely in sufficient depth to predict the effects of all his actions.

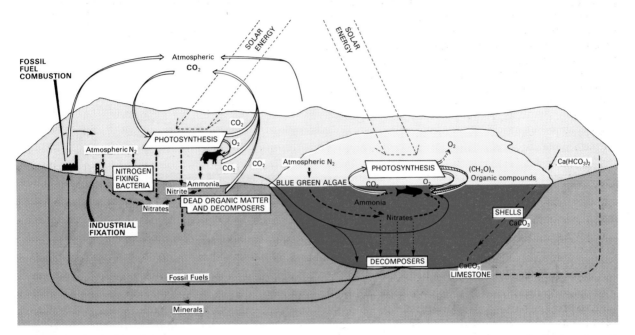

5.29   Cycles in the biosphere

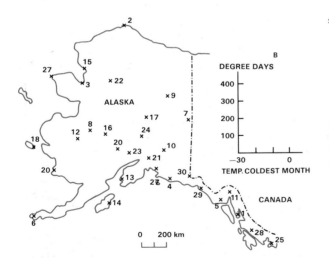

STATION	DEGREE DAYS*	TEMP. OF COLDEST MONTH	VEGETA-TION**	STATION	DEGREE DAYS*	TEMP. OF COLDEST MONTH	VEGETA-TION**
1	168	−5	S	16	306	−23	W
2	0	−28	T	17	411	−22	W
3	50	−23	T	18	0	−12	T
4	239	2	S	19	52	−10	T
5	89	0	S	20	36	−16	T
6	36	−2	T	21	71	−14	T
7	349	−25	W	22	218	−20	W
8	240	−19	W	23	319	−14	W
9	431	−28	W	24	41	−17	T
10	225	−21	W	25	364	2	S
11	327	−5	S	26	83	−7	S
12	280	−16.5	W	27	0	0	T
13	95	−5	S	28	366	−1	S
14	153	−1	S	29	95	−3	S
15	57	−29	T	30	28	−13	T

* Degree days above 10°C [eg if station had monthly means of 12° July and 11° August and below 10°C for other months D.D. = 2 x 31 days for July plus 1 x 31 days for August, total 93.]
** Major vegetation type S = sitka spruce hemlock forest, W = white spruce birch forest, T = tundra

5.30   Alaska

Review Exercises

1. Describe and explain the variations in soils at the world scale.
2. Review the factors responsible for the distribution and character of the soils of north Somerset.
3. Describe the various nutrient cycling models shown in this chapter. What do they tell us about the inter-relationships between climate, soils and vegetation in different environments?
4. With reference to specific examples discuss why forest gives way to grassland.
5. Describe man's modification of two natural eco-systems.
6. Figure 5.30 shows the vegetation and climate for thirty Alaskan stations whose location is shown on the map. Using a scatter graph (format shown at B) analyse and describe the climatic controls on the three vegetation types. Using the location map and an atlas describe the pattern of vegetation and comment on the effects of latitude, altitude and other determinants of climate in influencing the distribution.
7. Where and under what conditions does accelerated soil erosion occur? What land-use practices reduce its effects?
8. Explain the effects of increasing altitude on vegetation within the tropics.
9. How are plants adapted to excessive moisture, drought, salinity and cold?
10. Assess the relationship between climate, vegetation, soils and man in the Mediterranean Basin.
11. What are the ecological implications of mechanised, energy-intensive agriculture?

Further Reading

The following contain information on *field techniques* and themes for investigation:

Courtney, F. M., & Trudgill, S. T., 1976, *The Soil: An Introduction to Soil Study in Britain*, Arnold.

Hanwell, J. D., & Newson, M. D., 1973, *Techniques in Physical Geography*, Macmillan.

I Reading

Bridges, E. M., 1970, *World Soils*, Cambridge University Press.

Brady, N. C., 1974, *The Nature and Properties of Soils*, Macmillan.

Eyre, S. R., 1975, *Vegetation and Soils*, Arnold.

Findlay, D. C., 1965, *The Soils of the Mendip Hills of Somerset*, Soil Survey of G. B. Memoir.

Freeman, W. H., 1970, 'The Biosphere', *Scientific American*.

Nelson, J. G., & Chambers, M. J. (Eds.), 1969, *Vegetation, Soils and Wildlife*, Methuen.

Odum, H. T., 1971, *Fundamentals of Ecology*, Saunders.

Richards, P. W., 1973, 'The Tropical Rain Forest', *Scientific American*, December, 59–67.

Simmons, I. G., 1974, *The Ecology of Natural Resources*, Arnold.

Smith, P. J., 1972, *The Prairie Provinces*, University of Toronto Press.

Tivy, J., 1977, *Biogeography: A Study of Plants in the Ecosphere*, Oliver & Boyd.

Curtis, L. F., Courtney, F. M., & Trudgill, S., 1976, *Soils in British Isles*, Longman.

6 · Climate and Environmental Change: the Pleistocene and its Legacy

THE CHAPTER'S THEME AND PURPOSE

The world in August 17900 BP, with a mantle of snow and ice over large areas and with sea level 90 m lower, was radically different from today (Fig. 6.1). The effects of this kind of climatic change during and since the Pleistocene forms the theme of this chapter. It begins with an examination of the work of ice in the mountains and lowlands. The weight of ice sheets and the changing volumes of the oceans caused a series of land and sea-level changes. The nature and extent of this glacial isostasy-eustasy and its effects on shorelines and rivers forms the focus of the next section. This is followed by an investigation of the world of underground ice in the periglacial zone. The chapter concludes with a study of the legacy of the Pleistocene in the arid and semi-arid tropical world.

The Pleistocene glaciation: some possible causes

Geological evidence suggests the existence of four

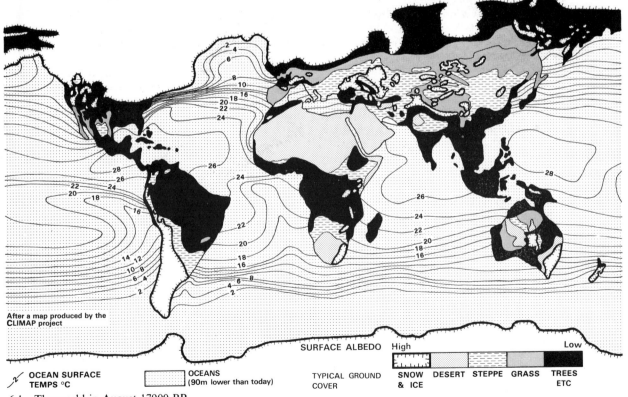

After a map produced by the CLIMAP project

SURFACE ALBEDO   High ▭▭▭▭▭▭ Low

| OCEAN SURFACE TEMPS °C | OCEANS (90m lower than today) | TYPICAL GROUND COVER | SNOW & ICE | DESERT | STEPPE | GRASS | TREES ETC |

6.1   The world in August 17900 BP

glacial epochs during the last 1,000,000,000 years. With the birth, movement and destruction of plates and with the continual erosion of the land only the most recent need really concern us here. For most of this period of 1,000,000,000 years the average global temperature was 22°C and even the poles were ice-free. About 50,000,000 years ago the earth began cooling and this trend culminated about 2,000,000 years ago with the onset of the Pleistocene glaciation. As Fig. 6.2 shows since then, apart from brief inter-glacials, snow and ice have been a dynamic element on the earth.

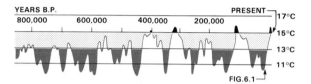

6.2 Global average air temperatures

The build-up of large ice sheets is a relatively slow process. Recent evidence suggests, however, that climatic change itself can be much more rapid. The evidence for this comes from a range of sources. Historical accounts can be used for the immediate past and in particular locations these can be extended back for several thousand years. Techniques such as pollen analysis and dendrochronology can extend our insights back further to the most recent deglaciation. Pollen analysis, for example, involves the collection of wind-blown pollen from peat bogs and it allows the reconstruction of earlier vegetation. Since some plants (see Exercise 6 on page 138) have well-defined tolerances to temperatures and precipitation this allows the reconstruction of past climates. Dendrochronology involves the identification of climatic changes from annual growth rings in trees. In the south-west of the USA the bristlecombe pine has allowed the construction of records going back some 8,200 years.

For earlier periods drill cores from the Greenland and Antarctic ice caps and from sea-floor sediments are used. In connection with the latter it has been discovered that a plankton—foraminifera—has a shell which coils to the left in water below 7°C and to the right above. This has been useful in discovering and dating a very sharp fall in ocean temperatures about 3,000,000 years ago. This technique is complemented by oxygen isotope measurements which involve determining the ratio between O^{16} and O^{18} isotopes in the foraminifera shells. High evaporation from the oceans—and hence snowfall onto the ice caps—removes O^{16} leaving greater proportions of O^{18}. In the case of glacial ice the ratios of oxygen isotopes at various depths in the ice caps have given a record of climatic conditions extending back over 100,000 years.

Such techniques, and the mass of geomorphological evidence described later in the chapter, give us a good idea of the magnitude and extent of climatic changes. But what of the causes? Climate reflects the flow of energy onto and away from the rotating sphere of the earth. It is influenced by a host of characteristics of the earth itself and of course by the sun. Before discussing some factors it is important to state that no *single* explanation of climatic change is acceptable.

The first question might be: Does the sun's energy-output vary? There is evidence that it does with changes in sunspot activity, and cycles of eleven and twenty-two years have been identified. The shape of the earth's orbit as it swings through space, its tilt and the precession of its axis (wobbles), also produces variations in solar radiation receipt. As long ago as 1920 Milankovitch on this basis postulated cycles of 20,000, 40,000 and 95,000 years. The latter is close to the 100,000 year pulse of the ice age in which we still live (Fig. 6.2). However, as the earth has experienced periods of 250,000,000 years without glaciations it is clearly not the only answer.

The surface of the earth and the atmosphere itself also need to be taken into account. One of the most dramatic changes in atmospheric composition is the injection of volcanic dust. The eruption of Tambora in the spring of 1815 in Indonesia has been estimated to have injected up to 200 km³ of dust into the atmosphere, turning world sunsets red and reducing temperatures in the Alps by over 1°C. Although individual events have short-lived effects, the last 2,000,000 years have seen a world-wide surge of volcanism. Plate movements too may be an important background factor by changing the arrangements of oceans (and hence ocean currents) and the relief of the continents. Raising relief barriers (the Himalayas rose 3,000 m between the Pliocene and early Pleistocene, for example) creates 'weirs' which upset the planetary wind systems and the patterns of temperatures and precipitation. However, it is important to remember that plate movements during the Pleistocene are unlikely to have exceeded 120 km so we have to keep this

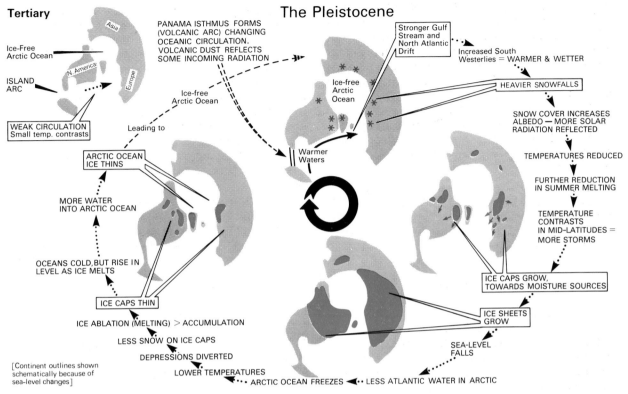

Tertiary

Ice-Free
Arctic Ocean

ISLAND
ARC

WEAK CIRCULATION
Small temp. contrasts

The Pleistocene

PANAMA ISTHMUS FORMS
(VOLCANIC ARC) CHANGING
OCEANIC CIRCULATION.
VOLCANIC DUST REFLECTS
SOME INCOMING RADIATION

Stronger Gulf
Stream and
North Atlantic
Drift

Increased South
Westerlies = WARMER & WETTER

HEAVIER SNOWFALLS

SNOW COVER INCREASES
ALBEDO — MORE SOLAR
RADIATION REFLECTED

TEMPERATURES REDUCED

FURTHER REDUCTION
IN SUMMER MELTING

TEMPERATURE
CONTRASTS
IN MID-LATITUDES =
MORE STORMS

ICE CAPS GROW,
TOWARDS MOISTURE SOURCES

ICE SHEETS
GROW

SEA-LEVEL
FALLS

Ice-free
Arctic Ocean

Ice-free
Arctic
Ocean

Warmer
Waters

Leading to

ARCTIC OCEAN
ICE THINS

MORE WATER
INTO ARCTIC OCEAN

OCEANS COLD, BUT RISE IN
LEVEL AS ICE MELTS

ICE CAPS THIN

ICE ABLATION (MELTING) > ACCUMULATION

LESS SNOW ON ICE CAPS

DEPRESSIONS DIVERTED

LOWER TEMPERATURES

[Continent outlines shown
schematically because of
sea-level changes]

ARCTIC OCEAN FREEZES ◄··· LESS ATLANTIC WATER IN ARCTIC

6.3   The dynamics of multiple glaciations

in context and take a broader time perspective.

One illustration of the interaction of factors is shown in Fig. 6.3 which is based on Ewing and Donne's theories. Beginning with an ice-free Arctic Ocean and no Panama isthmus, the sequence of changes leading into the cycle of glaciations is indicated in the annotation.

ICE IN THE MOUNTAINS

Snow accumulation

The climatic changes which began the Pleistocene glaciations should not be viewed simply as cooling. Lower overall temperatures are obviously important in encouraging snow accumulation at high latitude and altitude but such a build-up could reflect greater winter snowfalls or reduced summer melting or a combination of the two. Such climatic deterioration would be reflected in a lowering of the *snow line*, the altitude above which winter snowfall persists through the summer. Detailed terrain is also a factor. Shaded patches on slopes, those with a northerly aspect, for example, would have been favoured sites for snow to persist as would sheltered sites in the lee of a ridge where windblown snow would have accumulated.

Snowpatches produced by the interaction of climatic change and terrain factors would gradually etch themselves into the hillsides by a complex of weathering processes known as *nivation*. As the patch thickens and enlarges, the light fresh snow (density 0·06–0·55) becomes transformed by compaction, and successive melting and refreezing into a mass of firn (density 0·72–0·84). With further metamorphism firn is transformed into glacier ice. The interconnecting air passages become sealed, the crystals grow larger and the density rises to 0·85–0·91. This process of ice formation can take two hundred years in the cold and dry environments of the Greenland Ice Cap but is very much quicker in warmer and moister conditions.

Glacier budgets and movement

Glaciers range from small semicircular cirque glaciers a few hundred metres across to valley glaciers tens of kilometres long and hundreds of metres deep. With such diversity of size and shape it is useful to clearly *define* glaciers as bodies of ice and firn lying entirely or largely on land and which show evidence of either present or past movement. The effects of this movement will be looked at later, its causes form the focus of the next section.

Figure 6.4A is a model of a small glacier where the ice mass is viewed as a water *store* responding to *inputs* and *outputs*. Inputs consist of snow falling onto the glacier surface, together with snow which avalanches in from the surrounding slopes. The proportion of this which actually accumulates each year depends on the losses or outputs from the system. Some water is returned directly back to the atmosphere by evaporation from the snowpack. More obvious is the loss by melting. It is warmer at lower elevations, (*i.e.* to the right

of the diagram) and during summer the winter snow and some of the ice beneath it thaws. The resulting meltwater pours off the surface, at the edges, within and beneath the ice. The combined losses by melting and evaporation are known as *ablation*.

On the lower part of the glacier annual ablation exceeds accumulation, but as elevation increases a point is reached where the winter snowfall is not entirely melted—the snowline. Above this, accumulation exceeds ablation. The actual amount of net annual accumulation varies with environment. Glaciers in the Cascades of the western USA and in Norway may show accumulations of 4,000 mm (water equivalent) whilst in the cold desert areas of north Greenland it may not reach 10 mm a year.

Surveying of glaciers to establish their *budget* involves taking input and output measurements in both zones as well as assessing any changes in the volume of ice in storage (*i.e.* the glacier itself). Accumulation can be measured by digging pits or drilling cores. These show annual accumulation since each summer is marked by dirt bands caused by windblown dust and also in many cases layers of 'depth hoar' or partly melted and refrozen snow. Ablation can be determined by a combination of techniques measuring the downmelting of the ice surface itself or the amount of meltwater flowing away from the glacier.

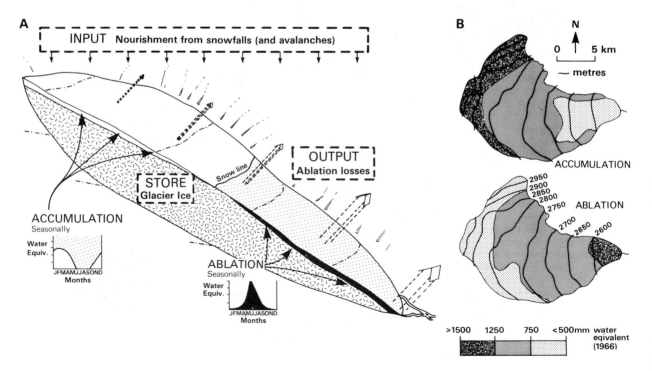

6.4A   Glacier model

6.4B   Ram River Glacier, Alberta

As one part of the glacier has more accumulation than ablation and the other more ablation than accumulation a transfer of ice under the influence of gravity will occur. When considering this transfer, or movement, it is important not to simply view glacier ice as the brittle substance familiar to us as the thin sheets covering ponds on winter mornings. When under the kind of pressures which exist within and beneath a glacier it can be plastically deformed. Glaciers move in a combination of ways. The ice may actually slip over the bedrock, a process accounting for up to 90% of movement in temperate glaciers although 50% is a more typical figure. Such slippage is an important process; without it the ability to erode would be much reduced. The ice may also slip on internal shear planes, a process likened to faulting in rock. Ice may also move by melting and refreezing or by a mass of intergranular adjustments.

By measuring *ice motion, i.e.* its velocity and direction, at different points on the surface and at depth, the overall pattern of movement can be established. This is shown in Fig. 6.5. The annotation emphasises the patterns of velocity and direction at different zones in the ice mass.

Changes in the bedslope profile and cross-sectional area of the valley may be reflected in variations in ice movement. Constrictions, or narrowings, of the valley produce zones of faster movement as do steeper segments of the long profile. In the latter case the upper zones of 'brittle' ice may fracture into a series of ice blocks, or *seracs*, separated by crevasses.

Seasonal variations in movement have also been detected. As the response of the glacier to changing inputs and outputs takes time (and this varies from glacier to glacier) the relationship is complex. The effects of one change, such as a heavier winter snowfall, may not have worked itself through the system before other changes begin operating. Using glaciers as long-term weather records is therefore potentially valuable, although our understanding of them is still limited.

Cirque glaciation

The snowpatches, mentioned earlier in connection

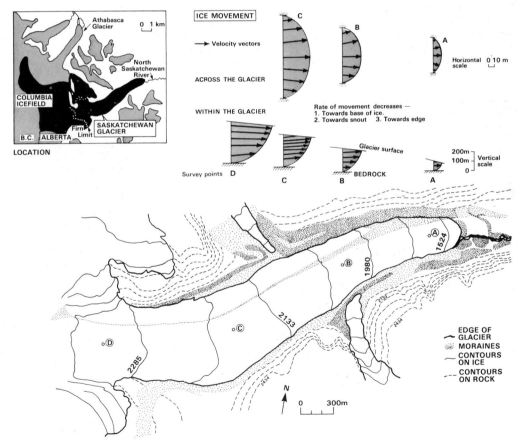

6.5   The Saskatchewan Glacier, Alberta

with climatic deterioration, gradually deepen their hollows by nivation with solifluction and meltwaters removing the weathered debris. As the patch enlarges it progressively bites back into the slope producing a small headwall. As the mass of firn and ice thickens the coefficient of friction is exceeded and movement begins—the birth of the cirque glacier.

Figure 6.6 shows a fully developed cirque glacier. The *cirque* is an amphitheatre-shaped hollow on the mountainside consisting of a headwall, concave floor and threshold (or lip). Above the ice frost-shattering etches the bare hillslopes, contributing a rain of angular debris which falls onto the glacier. Some of this may fall into crevasses in the ice and the remainder may become buried by falling snow. As the ice pulls away from the walls of the cirque under the influence of gravity it may pluck away rock which also becomes incorporated in the moving ice. This assorted rock material which ends up in the ice is known as *englacial moraine*. At the base of the moving ice it is used as tools to grind away the bed and sides of the hollow. This process of polishing and scraping is known as *abrasion* during which the bedload of the glacier is itself ground into smaller pieces.

The rotational movement of ice in the cirque glacier is shown in Fig. 6.6. In the ablation zone with the downwasting of ice the englacial material reappears on the surface and it finally accumulates at the snout as a *terminal moraine*. The larger boulders and pebbles are dumped here while the finer material can be carried away by the meltwater stream.

How the cirque glacier lengthens, widens and deepens its hollow has been a subject of some debate. Early views of frost shattering on the backwall at the base of the bergschrund are no longer accepted. This crevasse is not a universal feature of cirque glaciers and headwall heights exceed the depths of known bergschrunds. The microclimate within the crevasse would also produce only a few freeze–thaw cycles. Daily and seasonal flushings of meltwaters may, however, penetrate the back of the cirque, freeze and prise away rock fragments. Long-term climatic fluctuations might also produce a slow oscillation of the 0°C isotherm within the glacier. Such a freeze–thaw cycle, although long-term, could produce significant weathering over the whole glacial period. Melting and refreezing may also be produced by pressure variations, the weight of the ice on an irregularity in the bed could produce localised melting, rather like the

pressure exerted by an ice skater. The resulting film of water penetrates rock crevices and refreezes when the pressure is released, prising away a lump of rock which becomes incorporated in the moving ice. The removal of rock material by these freezing and thawing processes may cause pressure-release cracking (dilation) in the bedrock itself, providing sites for further freeze–thaw activity. Finally, of course, all this rock material can be used in the abrasion process mentioned earlier. The pattern of ice movement shown in Fig. 6.6 when combined with the processes described above allow the glacier to erode backwards into the hillside and in some cases to actually 'over-deepen' its bed when the slope rises towards the lip.

The production of the characteristic cirque shape may therefore reflect the operation of a number of processes. There is a lithological consideration too: bedrock susceptible to frost-shattering yet strong enough to maintain the steep headwall and sides is obviously required. In addition snowfall has to be sufficient to nourish the cirque glacier yet not so large that an ice cap accumulates and completely mantles the upland.

Where two cirques develop side by side, or back to back on either side of a divide—a sharp frost-shattered ridge or *arête* separates them. Where cirques develop on all sides of a mountain a pyramidal peak is

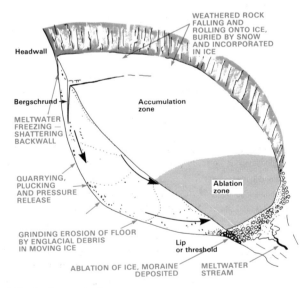

6.6   A cirque glacier

produced by the backwearing of the glaciers.

Valley glaciers

Where a number of cirque glaciers, or a high-altitude ice field (Fig. 6.5) feed into the line of a pre-existing river a valley glacier is formed. The surface features of the moving tongue of ice reflect the processes of erosion, transport and deposition associated with the glacial environment. Frost-shattered debris from the valley walls form a litter of debris at the edge of the glacier, the *lateral moraine*. Where two ice streams meet their laterals coalesce to form a *medial moraine* which snakes away down glacier. At the snout of the glacier where ice melts more rapidly than it moves forward, the assorted supraglacial, englacial and subglacial material is dumped as the *terminal moraine*.

The finer materials are moved away from the snout by meltwaters leaving the coarser rubble and boulders behind. Meltwater streams have pronounced seasonal and diurnal variations in discharge. At high discharges they can move quite large material but when discharge and velocity falls the meltwater streams have a much reduced competence and capacity. With such a variable hydrological environment, and a massive supply of loose debris of all sizes, the *outwash plain* down valley from the snout is a dynamic and changing environment. The spread of fluvioglacial material (known as the valley train) is reworked and crossed by braided streams. The finest material, the so-called rock flour, is carried in suspension, even at low discharges, and it gives the characteristic milky appearance to the meltwater streams.

The processes at work beneath the ice have been described earlier in connection with the cirque glacier. The moving ice ploughs up the pre-glacially weathered rock and soil, plucks and quarries fresh rock from the bed and sides of the valley and polishes and abrades bedrock surfaces. The glacier moves more slowly and occupies a larger cross-sectional area than a stream discharging the same amount of water in a year. Measuring *erosion rates* is far from easy but expressed as rates of lowering figures ranging between 0·005 and 2·8 mm a year have been obtained, *i.e.* ten to twenty times that of fluvial erosion! As glaciation is a relatively recent episode, the freshness of its landform legacy might have easily misled us. Such measurements, however, confirm the potency of moving ice and meltwaters as agents of landscape change.

Highlands after the ice

When accumulation and ablation change, the volume of the glacier alters and with it the position of the snout. During deglaciation downwasting exceeds replacement and the ice margin recedes, leaving behind its *terminal moraine*. These usually form lobate ridges across the valley, often with steeper ice contact slopes

6.7 A valley glacier in Baffin Island. The accumulation zone is at the top of the photo. The surface of the descending tongue of ice is heavily crevassed and medial moraines can be seen running along the glacier. At its sides are pronounced lateral moraines. The present ice snout has receded several hundred metres upvalley from a large terminal moraine. Meltwater streams have broken through the ridges of this moraine and are depositing a fan of outwash material.

(*i.e.* proximal slopes) and gentler distal slopes (*i.e.* facing away from the ice). They frequently consist of masses of unsorted unstratified angular debris, often with faceted boulders which have been dragged against the valley floor by the moving ice.

A terminal moraine of appreciable size reflects a stable ice-margin position. When the snout retreats each season may produce a smaller cross-valley moraine and a series of *recessional* (or washboard) *moraines* may be formed. When the ice margin becomes stable again for a period of years another terminal moraine is formed. If climatic fluctuations produce a re-advance of the ice the material littering the valley floor may be pushed up. Such *push moraines* frequently have gentle proximal slopes and steeper distal slopes of coarser less-compacted material.

When the ice finally melts it leaves a complex of landforms in the glaciated valley. At the larger scale the pre-glacial river valley has been straightened, deepened and widened by the ice. Its *cross section* resembles a catenary curve, frequently overdramatically described as a deep U shape. The interlocking spurs of the original valley have been planed off by the ice-producing *truncated spurs*. As a result of the deepening and widening of the glacial trough, tributary valleys no longer grade smoothly into the main valley, their lower portions have been removed and they hang above the trough. Streams draining such *hanging valleys* may join the main valley in a waterfall. Such straightening, deepening and widening stand as

mute testimony to the power of moving ice, its englacial debris and of course the recency of glaciation (Fig. 6.8).

The long profile of the valley may also show irregularities. Unlike a river valley the glaciated valley may have a steep *trough end* where cirques and ice fields fed into the main glacier, an increase in ice mass leading to increased erosive capability. The long profile may also be stepped, in contrast to fluvial valleys with their smoothly declining long profiles. The origin of such steps, or *rock bars*, has caused debate and remembering the principle of equifinality of form (different processes producing the same shape) no single theory of their origin is likely. In some locations well-jointed bedrock may have been particularly susceptible to plucking and removal, wider joint spacing being associated with the step. The steps may also reflect the joining of tributary ice streams where an increase in volume led to more erosion. In some cases they may reflect a narrowing of the valley where increased velocity led to erosion and deepening. Finally they may mirror irregularities in the original valley profile which have been perpetuated and accentuated by ice action.

The presence of *lakes* is a particular manifestation of this long profile irregularity. Elongated ribbon lakes may reflect the existence of rock basins scoured by the ice and/or the ponding back of water behind terminal moraines. They are essentially temporary features being infilled from their heads, and sometimes

6.8   A glaciated trough, the South Saskatchewan 10 km downvalley from the present glacier (see Fig. 6.5)

sides, by deposits carried by post-glacial streams.

At a smaller scale within the valley a range of erosional and depositional landforms are found. Erosional features include bedrock surfaces scarred with fine grooves or scratches known as *striations*. *Quarried* or plucked surfaces are also common, particularly in well-jointed rock. The *roche moutonnée* is a small-scale example of this process. Its 'upstream' surface has been smoothed by englacial debris as the ice streamed over and around the obstruction. As the ice slid past the obstruction the pressure melting effect ceased, water refroze in joints and the moving ice plucked away blocks of rock from the lee side.

A host of depositional forms may exist, too. Terminal and recessional moraines form ridges across the valley. The floor of the trough may also be littered with irregular hummocks and mounds of debris deposited and moulded by the moving ice. The melting and decay of stagnant ice (dead ice) is associated with a complex collection of glacial and fluvioglacial deposits. As it melts the decaying ice can leave the valley floor littered with the debris which it contained. This is known as *ablation till*, material lowered into position as the ice which carried it melts. This ablation till covers the unsorted *lodgement tills* which were deposited earlier by actively moving ice.

The role of running water in the deglaciation environment should not be forgotten. Meltwaters pouring off and beneath the ice can erode *drainage channels* flowing indiscriminately across ice, moraine and bedrock. As the ice melts and moraines slump only the portions of the channels carved in bedrock persist, often as 'in and out' loops on the valley sides. Meltwater also fills crevasses in the ice with sorted and rounded water-washed sands and gravels. Lenses of ice detached from the glacier become surrounded and

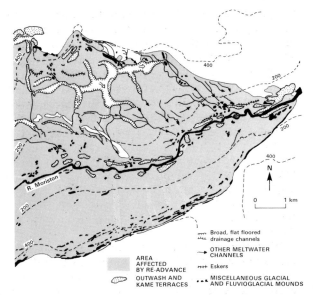

6.10 Fluvioglacial features, Glen Moriston, near Inverness

covered by various types of deposits. When this non-moving or 'dead' ice finally melts a hollow is produced which may fill with water. Such depressions, or *kettles*, reach their best development when the larger continental ice sheets decay. In this rather chaotic and changing environment small lakes and ponds frequently form into which water-washed clays, sands and gravels are sluiced. Where meltwater streams deposit material in contact with the ice itself *kame* deposits are produced. These may be in the form of 'terraces', marking the former course of streams flowing at the edge of the glacier, or they may represent deltas where meltwater streams flowing off the ice poured into water bodies.

If the above account sounds confused it does at least convey the complexity of a decaying glacier. Generally speaking material deposited by ice is unstratified. The water-transported and deposited materials, on the other hand, are frequently rounded and reworked and often sorted into different sizes and deposited in layers (*i.e. stratified*). The operation of the processes described above means that the glaciated valley is frequently veneered with *fluvioglacial deposits* which lie irregularly on the underlying glacial deposits. These often plug the floor of the trough.

When the ice finally disappears weathering, mass

6.9 Glacial erosion: a roche moutonnée, Cwm Idwal

wasting and fluvial processes begin to modify the landscape. The ribbon lakes and the tarns impounded behind rock lips or moraines in the cirques begin filling by sedimentation and vegetation encroachment. The bare rock walls as they weather produce a mass of scree which mantles their lower parts and modifies the form of the troughs and cirques. Weathering and mass wasting processes reshape the various erosional and depositional landforms. Finally the post-glacial streams begin carving into and reworking the valley train deposits frequently producing their own flood plains bordered by higher terraces of fluvioglacial material. Rock bars and moraines lying across the valley become breached and the progressive smoothing of the humid fluvial environment begins.

6.11 Small cirques, one with a moraine-dammed lake, on northeast flank of the Brecon Beacons above the Usk valley (also Fig. 6.13)

ICE IN THE LOWLANDS

Piedmont glaciers and meltwaters in the Welsh borderland

When glaciers enter a plain the restraining influence of the valley sides disappears and the ice spreads in a lobe across the lowland forming a *piedmont*, or mountain foot, glacier. The Wye basin has been chosen for examination here as its topography is familiar from earlier chapters. More significantly, however, it is an area where the landforms of piedmont glaciation are combined with the erosional and depositional effects of glacial meltwaters.

A reconstruction of conditions at the height of the last glaciation is shown in Fig. 6.12. Glaciation began on the highland plateaus of central Wales and sent ice streaming down into the Newbridge depression. The narrow section of the valley to the east of Builth (1 on Fig. 6.12) was ultimately overridden by ice moving south-eastwards. This 'piled up' against the northwest face of the ice-free Black Mountain massif, eventually to an altitude of 375 m. Most of this ice was diverted towards the north-east, entered the Hereford plain and spread eastwards as a vast piedmont glacier lobe. The thinning margins of this lobe nudged against the hills and ridges of the plain so that the exact position of the ice front was influenced by the pre-glacial relief. The remainder of the upper Wye ice moved southwards from the Black Mountains, over-topping the pre-glacial divide and flowed into the Usk valley where it joined ice from the Mynydd Eppynt, Brecon Beacons and Fforest Fawr. This ice stream, the Usk glacier, confined by the valley sides, reached to the south of Abergavenny (2) and also penetrated into the Vale of Ewyas (3). Ice flows upslope for short distances where there is a sufficient gradient on the ice surface and points 4, 5 and 6 are examples of such behaviour. Where the ice crosses the pre-existing divide in this way it is known as glacial *diffluence*.

What is the evidence for this reconstruction and the deglaciation which followed it? Some of the landform evidence is shown in Fig. 6.13. The maximum ice advance is marked by a series of terminal *moraines*. Where these are in well-defined valleys the moraine may form a distinct cross-valley ridge, as is the case with the 40 m high moraine at Llanfihangel Crucorney (A). In the case of the Wye lobe itself the moraine is more subdued, an irregular feature forming a belt of

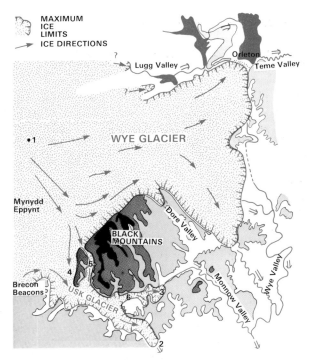

6.12 The central Welsh Borderland at the height of the last glaciation

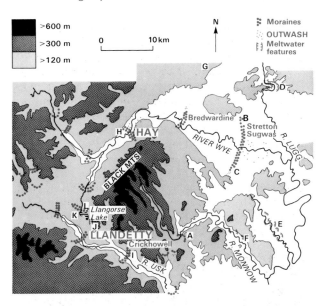

6.13 The legacy of glaciation

gentle mounds across the Hereford plain (B to C). Gravel pits in this moraine at Stretton Sugwas have exposed deposits deformed and pushed by the moving ice. Meltwaters flowing away from the ice margin have rounded and sorted this material into spreads of fluvioglacial deposits. Post-glacial stream action has cut into these deposits leaving gravel-veneered *terraces* a few metres above the current flood plain (evident from Fig. 4.64). Meltwaters also enlarged some river channels, for example at points D, E and F, so that the post-glacial streams appear *underfit*, or too small to have produced the large wavelength valley meanders. The production of such valley forms would require higher channel-forming discharges than exist today. The concentration of yearly runoff in the spring melt and reduced evapotranspiration losses in the cool climate with its scanty tundra-like vegetation and impermeable frozen ground would have combined to give much higher stream discharges in such zones bordering the ice.

In the lowland environment of the Wye valley the

erosive power of the ice has been overshadowed by its legacy of deposition. The piedmont lobe has mantled the surface with a layer of glacial deposits, most of which is local in origin and composed of Devonian material. However, occasional *erratics* are found. These are rocks distinct from the local underlying bedrock, Silurian rocks from the west (*i.e.* the Hanter Hill gabbro) being a case in point. The existence of erratics is evidence of the existence of some kind of erosional and transportational agency moving from west to east.

The stones within till sheets may also display a preferred orientation: being plastered into position beneath moving ice their long axes become aligned parallel to the direction of ice movement. *Till fabric analysis* is a technique of measuring the orientation of the long axes of stones in till sheets. Figure 6.14 is an example of a fabric analysis taken at point G on the northern side of the Wye glacier. Its preferred orientation confirms ice movement from the west.

In some situations which are not clearly understood the moving ice may mould the subglacial deposits into *drumlins*. These are elongated, smoothed egg-shaped mounds whose blunter (stoss) end faces the direction of ice advance. A belt of drumlins lies close to the northern flank of the piedmont ice lobe between Kington and Orleton. This gives rise to hummocky

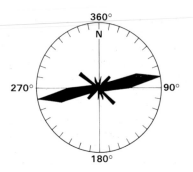

6.14   Till fabric analysis

terrain, with a local relief of about ten metres and frequent small ponds in the poorly drained sites between the drumlins (Figs. 6.15 and 6.16).

Figure 6.13 also shows the location of other moraines marking halts and minor readvances during deglaciation. These vary in size from the fifty metre high Hay moraine (H) to smaller arcuate gravel fans such as the Crickhowell moraine (I). As the ice thinned, more of the area became ice-free and the diffluent flow across divides ceased. When this occurred in the upper Llynfi valley the Wye ice filling the lower northern part of the valley caused the formation of a *proglacial lake*, *i.e.* a lake dammed up by the ice blocking the normal exit from the valley. This lake drained southwards over low points in the divide between the

6.15   Drumlins. Moulded drift from the Wye Glacier (point B on Fig. 6.16)

Llynfi and Usk valleys (J and K). As the ice disappeared from the Llynfi valley northwards drainage resumed although a shrunken Llangorse Lake remains impounded behind morainic barriers (L).

Ice frequently behaves in this way, temporarily blocking drainage lines and creating ice-dammed lakes. Water drains from such lakes across the lowest

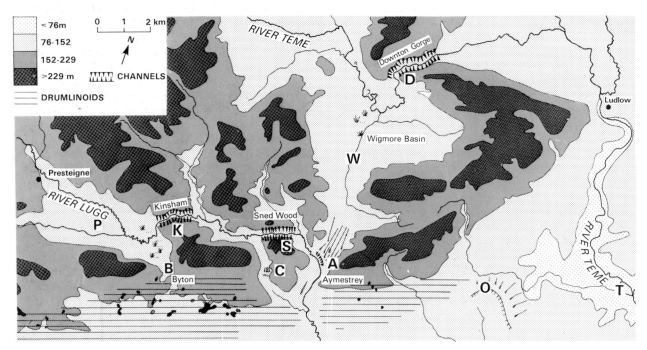

6.16   Meltwater effects in northern Herefordshire

point in the surrounding ice-free divide, as was mentioned above in the case of the ancestor to Llangorse Lake. The meltwater may also drain under the ice or along the glacier margins.

Figure 6.12 showed the existence of three such ice-dammed lakes to the north of the Wye lobe. This area is enlarged in Fig. 6.16, which shows that the Teme and Lugg rivers behave in rather unusual ways. Flowing from the west they enter basins (W and P in Fig. 6.16), ignore fairly pronounced exits (A and B) and proceed to plunge into steep-walled gorges (D and K). Both rivers in fact appear to have been *diverted* from their pre-glacial courses. The Lugg originally drained south-eastwards through the gap at B and the Teme southwards at A. How did this happen?

The maximum extent of the Wye ice is marked by a low (15 m) moraine at Orleton (O) and a belt of drumlinoid terrain stretches eastwards from this along the flank of the hills which formed the northern border of the ice lobe. The ice and moraine effectively blocked the preglacial drainage lines. Meltwaters filled the basin at P. Overtopping the ice-free divide, it began cutting a channel at K (Kinsham). The eastward-flowing water then utilised the line of an existing valley but with its exit blocked at C it carved a further channel at S (Sned Wood). Both these channels are still steep-sided and rock-walled in parts. Unusually for drainage channels they are relatively straight as they have exploited lines of structural weakness on the southern flank of the Ludlow anticline.

Water from the glacial Lugg arriving at point A (Aymestrey) found a further meltwater lake where the Teme had been dammed. This second lake overflowed at D (Downton Gorge) but the combined meltwaters still found a southern route blocked by the moraine at O and were again diverted eastwards (T). When the ice disappeared the meltwater valley incision had been so successful and the morainic plugs remained so effective that the rivers have remained diverted, although the post-glacial Lugg has breached the debris plug at A and resumed a southerly course.

This kind of evolutionary account of meltwater lakes and drainage diversions often sounds quite plausible but it is fair to ask about the evidence. Trying to explain the river courses without invoking ice-damming is far from easy and of course as Fig. 6.16 (and 6.13) shows there is ample evidence for the existence of the Wye glacier itself. The lakes themselves present more of a problem. In some situations *lake shorelines*

may have been produced, glacial Lake Llangorse shows some evidence of these at 192 m along its eastern side about 36 m above present lake level. In many cases however, there was neither the stability of water level, supply of beach materials nor the length of open water season for wave action to produce beaches. However, material washed into and settling out in such water bodies can provide useful evidence, the fan of material spreading north from A (in Fig. 6.16) is a case in point. In some locations too varved clays may be found. *Varves* are seasonal lacustrine sediments. The summer snowmelt streams vigorously transported clays, sands and gravels, dumping them into the lakes. The coarse sediments settled out first and later during the winter, when the lake was calm and covered with a skin of ice, the finest clays settled out. As the seasons progress and the years pass the lake floor becomes covered with such alternating layers. Such varved clays have been reported from basin P.

The existence and significance of meltwater lakes and the influence of glaciation on drainage diversions is now well accepted. In the further readings at the end of the chapter examples will be found of other lakes and diversions in Britain.

PLEISTOCENE CHRONOLOGY

So far no mention has been made of multiple glaciations although Fig. 6.2 showed that temperature conditions fluctuated during the Pleistocene. Figure 6.10 showed the effects of a readvance, a cooling pulse superimposed on the general trend of deglaciation. The techniques involved in producing a detailed time sequence, or chronology, are beyond our scope and as the most recent ice advance tended to destroy the landform legacy of the earlier ones our focus is on the most recent glaciation. An additional problem is that although distinct glacial periods have been recognised in various parts of the world they may not have peaked at the same time nor lasted for the same period.

In the case of the Wye glacier described above the conditions have been attributed to the last major glaciation, termed the Devensian. In central Europe the most recent glacial period is known as the Würm, in the Baltic as the Weischel and in North America as the Wisconsin. Figure 6.17 gives the terminology of the Pleistocene used in these three areas and Great Britain.

ICE OVER ALL: CONTINENTAL GLACIATION

The Wisconsin glaciation of North America

Our attention now shifts from the small cirque, valley and piedmont glaciers to the vast ice sheets which were such a dramatic feature of the Pleistocene. Imagine a climatic deterioration occurring in North America—snow lines would lower, the small ice caps would expand and outlet glaciers from them flow onto lower ground. New glaciers would appear in upland areas too warm at present to nourish them. With the passage of time ice would thicken, coalesce and spread. The growing ice sheet would be best nourished by the heavier snowfalls occurring on its southern and maritime margins. After several thousands of years ice may have built up to form a massive *continental ice sheet* similar to that shown in Fig. 6.18. The concept of such massive ice sheets repeatedly overwhleming large continental areas, with life returning as the vast silent immensity of inland ice decayed, now seems obvious. Yet the first attempts to explain glacial erratics and moraines involved notions of icebergs floating in the massive seas of the Biblical flood! The term drift which is applied as a general term to glacial deposits dates back to this period. During the nineteenth century the

Great Britain	C. Europe	Baltic	N. America
DEVENSIAN	WURM	WEISCHEL	WISCONSIN 67000
Ipswichian	*Uznach*	*Eemian*	*Sangamon*
Wolstonian	Riss 128-225000	Saale	Illinoian 128-180000
Hoxnian		Holstein	*Yarmouth*
Anglian	Mindel 350-600000	Elster	Kansan 230-300000
			Aftonian
	Gunz 800-1400000		Nebraskan 330-470000

Blue: Glacial periods   Italic black: Interglacials

6.17 Subdivisions of the Pleistocene

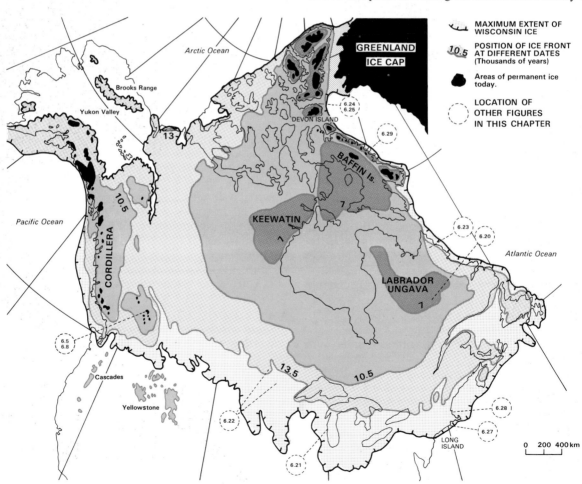

6.18 The Wisconsin ice sheet

awareness of continental glaciations slowly grew, the Swiss zoologist Louis Agassiz being the foremost protagonist. Although we may now question the causes and timing of these events the earlier debate is long since over.

Figure 6.18 shows a reconstruction of North America at the height of the last glaciation about 20,000 years ago. The ice sheet consisted of two parts—the Cordilleran ice in the west, which grew and accumulated in the Rockies; and in the east the Laurentide ice sheet which probably reached a thickness of 3–4 km. The volume of ice has been estimated at 12,000,000 km³. Remembering that large ice sheets existed in other areas of the world (see Fig. 6.1) the removal of such large amounts of water from the oceans and its storage in ice sheets produced a world-

wide lowering of sea-level. This enabled the Wisconsin ice to spread onto areas of the continental shelf. Figure 6.18 also shows the existence of large ice-free areas in northern Alaska, the Yukon and the land bridge connecting Asia and North America. Temperatures were certainly low enough for snow and ice accumulation so why did these northern areas remain ice free? If the relief is considered, the 5,000 m high barrier of the Alaska Range and the St. Elias Mountains can be seen to have cut off this area from the moist Pacific airmasses.

The decay and shrinkage of the Wisconsin Ice over the 8,000 years of the deglaciation phase is also shown in Fig. 6.18. As an agent of erosion, transport and deposition the ice has bequeathed a varied legacy of landforms, some of these are shown in Fig. 6.19 for

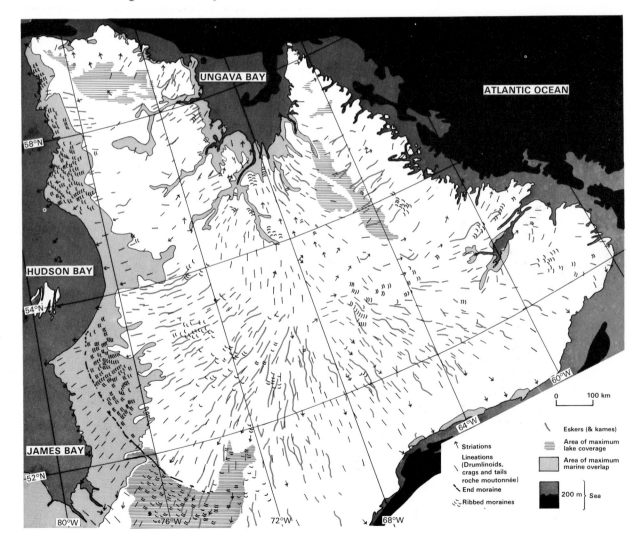

6.19   Glacial landforms in Labrador Ungava

the Labrador Ungava area.

The overall *erosional impact* of continental glaciation is difficult to assess. Unlike valley glaciers (which moved through valleys with steep gradients and with supplies of weathered debris from the ice-free hillsides) the continental ice flowing over slight relief, at the rate of a few metres a year, produced less striking changes in the landscape. As Bird wrote, 'the widely held view that the flatness of the Canadian Shield is due to glacial erosion is certainly untrue'. Its flatness results from long-term erosion over the previous periods of earth history. The effect of the ice sheets may have been to merely remove weathered bedrock and soil. The volume of glacial deposits and their mineralogical freshness does suggest, however, that some removal of bedrock did in fact take place. The stripping off of soil and regolith has left areas of polished and abraded bedrock. Striation orientations on these indicate the directions of ice movement. If you look at Fig. 6.19 you will see that the striations suggest a radial dispersal of ice from the centre of the peninsula.

Most of the northern and eastern areas overlain by Wisconsin ice consists of the Canadian Shield, 80% of which is composed of granitic-gneissic rock. Jointing lines, intersecting at right angles, were susceptible to quarrying and rock removal. Ice action in this situation produced *stoss and lee topography*, smooth gentle slopes and quarried lee surfaces. Between these rock knobs were basins where material had been removed along the jointing lines. Such basins are now frequently water filled and the lakes have an angular pattern reflecting this structural influence.

At the margins of the ice at its maximum extent the transported material accumulated as terminal moraines, the lobes of ice producing a series of conspicuous arcuate ridges beyond which meltwaters deposited outwash deposits. Behind the terminal moraines the pre-glacial relief was buried by a varied mantle of lodgement and ablation tills, which often lay on top of the till sheets of earlier glaciations. The thickness of the till appears to vary with the area's rock resistance, ranging from a metre or so on the harder rocks of eastern Canada, to ten metres or so over shales in southern Ontario to in excess of a hundred metres over soft unconsolidated sedimentaries in the Prairies.

As the ice decayed a range of glacial deposits were exposed, Fig. 6.21 maps a selection of these in a small area of Wisconsin. *Till plains* mentioned above cover the largest areas, varying from flat surfaces to undulating plains with a local relief of ten metres or so. *Drumlin swarms* may also occur, like the 20,000 in the drumlin field to the north of Lake Ontario. Where ice masses were abandoned to decay slowly a pitted surface of *hummocky moraine* results.

As the ice thinned and warmed at its margins meltwater flowing towards its edge formed streams flowing within and beneath the ice. Such streams would have carried rock material, rounding it and sorting it and depositing it both along their beds and where they

6.20   Glaciated shield terrain, Labrador. In this vertical photo the lakes appear black, woodland-covered tills as a stipple. There are areas of bare rock and an esker at E

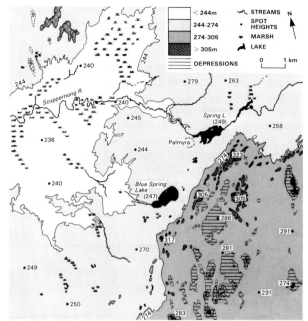

Running diagonally across the centre of the map is a till plain in the North west of which are three drumlins. In the South west is a terminal moraine beyond which is an outwash plain pitted by kettles.

6.21   Glacial deposition, Palmyra, Wisconsin

exited from the ice. The water could have flowed under hydrostatic pressure in the ice-walled tunnels and under such conditions natural syphons would allow the water to flow 'uphill' for limited distances. When the ice finally melted the stratified water-sorted sands and gravels would remain as *kame* and *esker* deposits. Eskers are long winding ridges of sands and gravels, the larger ones reach thirty metres or more in height and may extend over a hundred kilometres in length. The direction of meltwater flow would have been towards the thinning ice margin and thus at right angles to the ice front. The pattern of eskers shown in Fig. 6.19 is therefore consistent with a deglaciation involving the gradual retreat of the ice towards a small residual cap in the centre of the peninsula.

Figure 6.19 shows only one terminal moraine system but large areas of the smaller *recessional moraines*. These ribbed moraines are often only a metre high and 25–100 metres apart. Their origin is still disputed, the problem is that looking for analogies from the present-day Greenland ice cap doesn't necessarily tell us what processes operated when the 'warmer' and 'wetter' Wisconsin ice sheet decayed. These features could represent annual moraines formed by the squeezing out of waterlogged till from beneath the ice. The mapping of these moraines and the eskers, drumlins, moulded drift and other lineations has been possible with the aid of air photographs: the size of the Canadian North would have made the production of such maps as Fig. 6.19 impossible without this technique.

The role of *meltwater* at and beyond the edge of the ice sheet was also important. At its maximum the southern margin of the Wisconsin ice had few proglacial lakes, the general slope of the land to the south allowed meltwaters to drain away from the ice front. As the ice retreated the situation changed, there was more meltwater and land sloping towards the ice front became exposed so that a series of huge lakes were born. Figure 6.22 is an example of the largest of these, glacial Lake Agassiz. Born about 12,500 years ago in the Red River valley of North Dakota it initially drained southwards via the Lake Traverse outlet and the Minnesota River. During the following 5,000 years the lake level fluctuated at least thirty-five times as the position of the ice front changed; different outlets were used and the land rose after the weight of the ice sheet had been removed. Although its maximum area at any one stage was 200,000 km² it affected 500,000 km² stretching from the Dakotas to Saskatchewan. The beds of such lakes became covered with silts and clays and when water levels were stable for some time sandy beaches and deltas were deposited and low bluffs and benches were eroded.

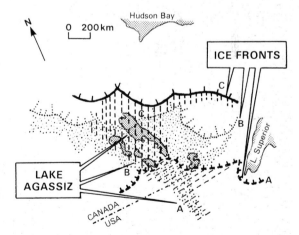

6.22   Lake Agassiz changing as the ice sheet retreats

Overflows from such lakes and meltwater from the continuing ablation of the ice sheet were prevented from draining towards Hudson Bay by the remaining ice. In the prairies this therefore led to the creation of meltwater channels or *spillways* draining south-eastwards. The water carved quite large valleys up to 70 m deep and a kilometre or so across. One of these was exploited in the South Saskatchewan Project mentioned on pages 108–11. Such spillways may now be dry or occupied only by small rivers. Their equivalent in the North German Plain, formed by meltwaters draining westwards to the south of the Scandinavian ice sheet, are known as *Urstromtaler*.

Interpreting the evidence

It is intended in this section to illustrate how late Pleistocene events have been reconstructed in two areas, north-eastern Labrador and Devon Island.

Figure 6.18 showed Labrador to have been over-ridden by ice at the Wisconsin maximum. However, there is some evidence that parts of the rim of coastal mountains, the Torngats, protruded through the edge of the ice sheets as *nunataks*. The evidence for this is detailed in the annotation on Fig. 6.23A.

As the ice sheet thinned the divide between the Atlantic and Ungava Bay became exposed but ice remaining to the east and in the north dammed up a series of glacial lakes. These were up to 320 km long and 80 km wide, their levels fluctuated as the position of the ice changed and the water escaped eastwards (Fig. 6.23B). Later, as the ice retreated, northward drainage was resumed and the lakes drained. The final remnants of the ice sheet decayed in the central part of the peninsula about 6,000 years ago. The interpretation of events in this way involves the identification of erosional and depositional features, careful mapping of their locations and elevations and finally attempting to establish a relative chronology, hopefully aided by absolute dating.

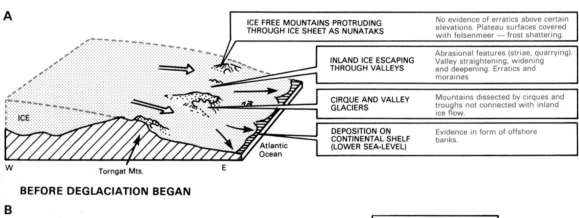

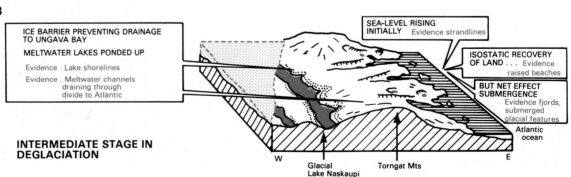

6.23   Deglaciation evidence in N.E. Labrador

The second example brings us to the question of changes in land and sea-levels. Mention has already been made of two relevant facts. Firstly that the growth of ice sheets led to world-wide sea-level lowering, and their melting to a sea-level rise, an *eustatic change*. Secondly the weight of the continental ice sheet depressed the crust so that when the ice melts and its weight is removed the crust 'rebounds' back, a process of *isostatic recovery*. The effects of this crustal warping, mentioned in the case of Lake Agassiz, are also evident in Labrador Ungava where the glacial lake shorelines are tilted, or rise, towards the centre of the peninsula. As the lake surfaces didn't slope, this tilt reflects the crustal recovery of the central part of the peninsula during the thinning of the ice and after its final disappearance.

At coasts which were glaciated or close to the ice margins this interaction is quite complex, because the two processes of eustatic rise and isostatic recovery operated at different rates. During the 1960's in

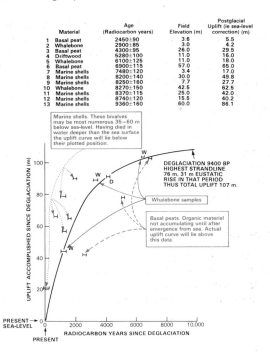

	Material	Age (Radiocarbon years)	Field Elevation (m)	Postglacial Uplift (ie sea-level correction) (m)
1	Basal peat	2450±90	3.6	5.5
2	Whalebone	2900±85	3.0	4.2
3	Basal peat	4300±95	26.0	29.5
4	Driftwood	5280±100	11.0	16.0
5	Whalebone	6100±125	11.0	18.0
6	Basal peat	6900±115	57.0	65.0
7	Marine shells	7480±120	3.4	17.0
8	Marine shells	8200±140	30.0	49.8
9	Marine shells	8250±160	7.7	27.7
10	Whalebone	8270±150	42.5	62.5
11	Marine shells	8370±115	25.0	42.0
12	Marine shells	8740±120	15.5	40.2
13	Marine shells	9360±160	60.0	86.1

Marine shells. These bivalves may be most numerous 35–60 m below sea-level. Having died in water deeper than the sea surface the uplift curve will lie below their plotted position.

DEGLACIATION 9400 BP HIGHEST STRANDLINE 76 m, 31 m EUSTATIC RISE IN THAT PERIOD THUS TOTAL UPLIFT 107 m.

Whalebone samples

Basal peats. Organic material not accumulating until after emergence from sea. Actual uplift curve will lie above this data.

6.25   Post-glacial uplift, Devon Island

6.24   Devon Island, a raised strandline

Canada researchers became interested in discovering and explaining the spatial and temporal pattern of these events. Immediately after a coastal area became ice-free a marine transgression occurred, the sea extending inland over large distances where the relief was gentle. As the crust began to recover the sea would retreat, if the rate of isostatic recovery was greater than the continuing rate of sea-level rise. The extent of this marine overlap is shown in Figs. 6.19 and 6.26 and the term *marine limit* is applied to the highest point reached by the sea.

Figure 6.24 shows a series of strandlines, low shingle ridges representing former beaches, on the north coast of Devon Island (Fig. 6.18 shows its location). The marine limit here lies 76 m above sea-level. If the height and age of the strandlines can be discovered a *post-glacial uplift curve* can be produced. Figure 6.25 lists the results of one investigation. The dates are C_{14} figures from radiocarbon analyses of organic material. The whalebone dates may be particularly reliable, the whales died after being stranded in shallow water, their massive skulls lying on the shingle as the land rose. Being heavy they were unlikely to be moved by scavenging animals or by wave action and the dense ear bones would resist contamination by younger organic matter which would give a false C_{14} date for the whale's death. By surveying the elevation of the beaches and dating them information on the extent and rate of uplift can be discovered. To find the total uplift eustatic changes have to be taken into account by obtaining the history of sea-levels from tectonically stable areas (*i.e.* those removed from the effects of glacial warping). If these corrected elevations are plotted on a graph (Fig. 6.25) the post-glacial uplift curve is produced. This shows the rapid initial uplift of 4 m/100 years with the rate of recovery slowing through time. Rates of up to 10 m/100 years have been discovered for other areas of northern Canada.

Figure 6.26 summarises some details of crustal warping in Canada. The shape of uplift curves suggests, and geophysical evidence such as gravity anomalies confirms, that recovery is not yet complete. Figure 6.26 also shows the current rates of movement and estimates of uplift still to be achieved. Although the Wisconsin ice disappeared 6,000 years ago its effects will continue to be felt well into the future.

THE EFFECTS OF CHANGING SEA-LEVELS

If Figs. 6.26 and 6.18 are compared maximum isostatic effects are found at the centre of the ice sheet. At the margins, where the ice was thinner and lighter the crustal depression (and later recovery) has been less, as can be seen from the lower elevations of the marine limit. Various estimates of eustatic lowering during the glacial periods have been produced, ranging up to 145 m below present sea-level. In areas unaffected by the continental ice sheets and their associated crustal warping the period since the glacial maxima has therefore been one of rising sea-levels. This marine *transgression* proceeded rapidly during deglaciation, but appreciably slowed about 6,000 years ago.

Determining current trends depends on obtaining long-term records, which are still rather scarce. Future sea-levels depend on the direction of climatic change.

6.26 Post-glacial crustal movement, Canada

The melting of the Antarctic Ice sheet could raise world sea-level by about 90 m, a return to a glacial period would lower it. The goal of predicting the direction, intensity and timing of global climatic change, of obvious use in connection with such short-term features as drought cycles, might thus be of more than passing interest to our descendants! One problem to add to the complexity of the earth's natural systems is man himself. By burning fossil fuels he is modifying the atmosphere and by clearing vegetation for croplands he modifies the earth's surface. He has introduced a 'wild card' into an already complex game.

Before examining some of the effects of climatic change there are two cautionary points. Firstly this discussion focuses on the repercussions of glaciation. It is important to remember, however, that the relative levels of land and sea may well be affected by faulting, volcanism and other earth movements. Similarly, tectonic forces and sedimentation may produce changes in the volume of the ocean basins and thus shoreline positions. Secondly, although the erosional, depositional and transportational work of waves will be mentioned a comprehensive treatment of coastal processes and forms will not be attempted.

Beyond the areas affected by glacially induced crustal warping a range of coastal features have been produced. Inundation of an unglaciated upland which had been dissected by fluvial action results in the drowning of the lower parts of the river valleys. Such inlets with their shallowing branching tributary arms are known as *rias*, common in south-west England. Where inundation occurs in an area with relief and structural trends running parallel to the line of the coast the short transverse and elongated longitudinal valleys will be flooded. This results in sea inlets parallel to the coastline and a series of elongated offshore islands and peninsulas. Such coastlines are known as *Dalmatian* coasts, named after the example on the eastern shore of the Adriatic.

Rising sea-levels in unglaciated lowland regions would cause widespread inundation. Broad shallow bays and branching inlets would mirror the relief of the drowned fluvial landscape. With shallow waters offshore and a supply of weathered debris from previous sub-aerial weathering such coastlines are quickly modified by marine action. Bars and spits

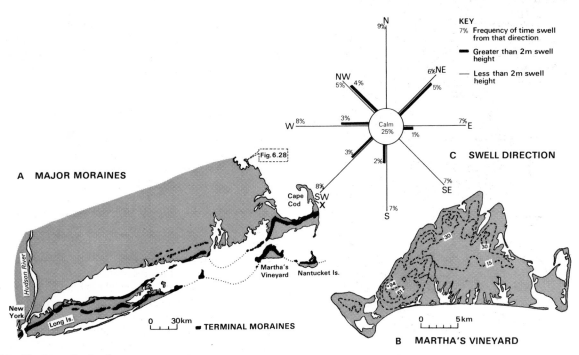

6.27   The New England coast

form and smooth the shoreline irregularities. In the lagoons behind the bars sedimentation proceeds, encouraged by vegetation encroachment on the tidal flats. An atlas map of the east coast of the USA shows (even at a large map scale) the effects of these processes on the Virginia and Carolina coast.

An identical range of processes operate when the lowland inundated is one mantled with glacial deposits. Figure 6.27A shows the terminal moraine systems which run through Long Island eastwards towards Cape Cod and Nantucket Island. If you look at the shape of the moraines you will notice that Martha's Vineyard (Fig. 6.27B) is in an interlobate area. The morainic arcs of the Cape Cod and Buzzards Bay ice lobes joined in the north of the island producing the 94 m high ridge which forms the backbone of the island. To the south originally stretched an outwash plain. With the return of water to the oceans the rising sea began working on this mass of unconsolidated material.

How did the waves achieve this? In the open ocean they are simply oscillations of water. In shallowing water this circular motion is hindered at its base, the leading edge of the crest steepens, water actually moves forward and the wave breaks. In this process particles of sand and gravel may be carried up the shore by the *swash* of the breaking wave. Some of the swash percolates into the beach. The *backwash* flowing down the surface of the beach between the arrival of the waves therefore involves less water than the swash. Sediment, which may have been picked up by the rapidly moving water in the surf zone, is therefore carried forward to build the beach. The slower backwash is unable to remove all the material back down the beach. Under storm conditions the swash may carry larger particles forward, throwing them up to form a storm ridge at the top of the beach. Lower down the beach, however, the larger and more frequent storm waves generate a powerful backwash, since the percolation rate into the beach has not increased. This backwash effectively combs material down the beach. After the storm it may be returned in the beach-building process described above. When the seabed is composed of a sloping mass of outwash sands and gravels this kind of wave action is able to quickly construct beaches.

A perfectly smooth plain is an over-simplification of the microtopography of an outwash plain. It may well be 'rilled' by a series of small valleys running

away from the ice front. If such a dissected outwash plain is inundated by the sea we have a miniature 'ria' type coastline. Being composed of fluvioglacial debris of course it responds to wave action far more rapidly than a rocky coast where erosion is slower and there is a much reduced supply of mobile sand and gravel. The tips of the ridges stretching out as miniature promontories are subjected to vigorous wave attack and consequently retreat. Waves are refracted on either side as they try to approach the promontory sides 'straight on'. Some of the material eroded from the tips is transported laterally and deposited in the bays. Gradually beaches build across the inlets from the promontories, the headland continues to be worn back and the coastline becomes smoothed in a line normal (at right angles) to the line of approach of the dominant waves. This process is visible on the south shore of Martha's Vineyard (Fig. 6.27B).

When waves approach a beach obliquely the swash proceeds up the beach diagonally but the backwash returns directly towards the sea. Particles carried or rolled by the water can thus move along the line of the shore in a kind of zig-zag path in the surf zone, moving diagonally up the beach in the swash and directly down the beach slope in the backwash. The effect of this *beach drifting* can be seen at a number of points around the shore of the island (see Fig. 6.27B) where spits have grown across bays and inlets. The process demands a plentiful supply of mobile material and waves which approach the shore diagonally. Both conditions are present in this area, the wave frequency rose (Fig. 6.27C) confirming the latter point.

The ability of a wave to erode and move material is influenced by its height. One control of this is *fetch*,

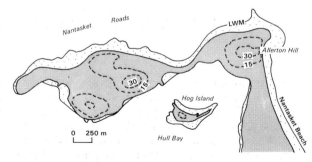

6.28   Drowned drumlins

A SECTION ALONG FJORD

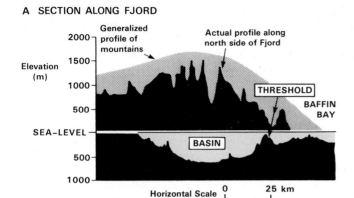

B COASTLINE PLAN AND SEA DEPTHS

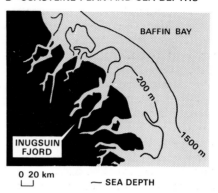

6.29   Inugsuin Fjord, Baffin Island

the greater the distance the wind can blow over the sea the larger the potential waves. Figure 6.27A shows that fetch varies from a few kilometres to the north of Martha's Vineyard to thousands of kilometres southwards across the open Atlantic. Does this have any effect on the apparent development of beaches and spits around the island?

In the case of Cape Cod the central part of the peninsula is retreating under the attack of Atlantic waves. Interestingly from point X (Fig. 6.27A and C) longshore drift moves material to both wings of the peninsula. Wave refraction around the northern tip (Cape Cod itself) has produced a pronounced recurved spit. Westwards from the Cape in the area shown in Fig. 6.28 a drumlin field has been drowned. The compact tills of the drumlins originally formed islands but marine processes have eroded parts of them and linked them together with beaches.

The processes described above operate on all coasts. The recency of transgression, the supplies of unconsolidated material and the power of Atlantic waves, however, combine to make the New England coast a dramatic example of the marine reworking of glacial deposits.

The fjord problem

Fjords are found where a mountainous area dissected by valley glaciation has been partially submerged. These long deep inlets display all the features of a glaciated trough. In many cases their plan view is rectilinear or angular, a feature attributed to the fault fracture system which had guided the pre-glacial river incision and which the glaciers have exploited.

Figure 6.29 represents the long profile of Inugsuin Fjord on Baffin Island in northern Canada. During the Wisconsin glaciation ice streamed eastwards, passing through the mountain rim towards the lower elevations in Baffin Bay. Similar escapes of ice produced fjorded coastlines in Labrador, British Columbia and Alaska and in Norway, Chile and New Zealand. Being close to the edge of the ice sheets, isostatic recovery

6.30   A raised beach, Wester Ross

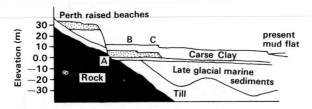

6.31   Diagrammatic section of the mid-Forth valley

was unable to compensate for the effectiveness of glacier scour in the valleys. The rock basins have therefore been flooded by the sea.

Two features of fjords have caused geomorphological debate—the *threshold* or shallowing at the mouth, and their relationship with the offshore submarine topography. The threshold of Inugsuin Fjord shallows to 150 m from the 600 m depths in the inner fjord. One clue to its origin is shown in Fig. 6.29—the threshold occurs where the ice emerged from the confines of the valley onto the lowland. The spreading and thinning ice was capable of less erosion and also began floating on the sea. Most thresholds are therefore bedrock features and not morainic in origin.

Seawards from the fjord mouths, troughs in the sea bed have sometimes been detected. In the case of our Baffin Island example there is a 900 m deep submarine canyon. In the light of what has been said earlier in the chapter you might consider if ice itself could be responsible. The most reasonable explanation is that they may have been eroded by *turbidity currents* which started life as sediment-laden meltwaters from the glacier snouts.

Sea-level fluctuations in Scotland

The marine transgression which occurred in arctic Canada produced few erosional features. Rapid uplift and short open water seasons in the summer, meant that waves usually had only the time and energy to produce constructional forms. In other locations with more wave energy and longer periods of sea/land stability marine erosion could produce wave-cut platforms and cliffs. With crustal warping these can now be found as raised platforms and beaches. In Scotland such *wave-cut features*—up to 800 m across and with cliffs as high as 60 m backing them—occur in many locations, especially on the west coast. They are generally higher (*i.e.* up to 45 m OD) close to the ice-sheet

centre which lay over Rannoch Moor, occurring today at progressively lower elevations away from this area. Glacial erosion and deposition has been reported on some of the platforms supporting the view that many of the higher ones may well have been formed during an interglacial pre-dating the last glaciation.

In eastern Scotland areas like the Forth valley have experienced a complex history of changing sea-levels and post-glacial tilting. Figure 6.31 summarises the work of Sissons where you can find a thorough discussion of this topic. The diagram shows the rock floor of the valley overlain by till. The main late glacial shoreline (A) was formed at a period of lower sea-level—after all there was still a lot of ice to melt! A period of marine transgression followed which ended at the time of the Loch Lomond readvance of the ice. After this stage falling sea-levels produced beaches B and C, dated to 9600 and 8800 BP respectively. In the mid-Forth valley these have been buried by the major transgression occurring after 8500 BP which covered the beaches with a layer of estuarine mud. Intermittent falls in sea-level occurred after this as can be seen from the profile of the upper surface of the clays. As post-glacial climate passed into warmer and wetter phases, extensive peat bogs (carse as they are known in this area) developed on the surface of the clays.

This example has been included to show the complexity of land and sea changes—the oldest beach isn't necessarily the highest! It also shows (and Sissons' discussion shows much better) how detailed survey and dating are required to explain these changes. In the case of the Canadian examples mentioned earlier this hasn't been possible. The marine limit is time transgressive. It is simply the highest point reached by marine overlapping and could have been formed at any time during the 8,000-year span of deglaciation. Unlike the Scottish examples, too, it doesn't represent the shoreline at a particular moment in time. Finally it should be noted that a section like Fig. 6.31 doesn't show the tilting of these shorelines which has occurred since they formed and were abandoned. The oldest generally show the greatest tilt away from the ice centre.

This section began with simple cases of inundation and has ended with complex examples of sea-level changes. The problems of coral reef and atoll origins and the detailed workings of the marine erosional and depositional systems have been omitted. A study of these would obviously complement the 'glacially' oriented views presented here.

ICE UNDERGROUND: THE PERIGLACIAL WORLD PAST AND PRESENT

Processes and forms

The term *periglacial* is applied to those areas which are marginal geographically and climatically to the ice sheets. They are environments of significant *frost action*. The products of this mechanical weathering are more varied in size than we sometimes think, ranging from particles of a micron in diameter to those as big as houses. The duration and intensity of frost action, together with the type of rock, affects the amount of weathering as well as the actual size of the material produced. Non-porous granites on the Canadian Shield have weathered at a rate of 3 mm/1,000 years whilst well-jointed limestones in the same area have mechanically weathered at a rate of 150 mm/1,000 years. The products of this frost action includes *felsenmeer* littering horizontal surfaces, apron-like spreads of scree lying beneath frost-riven cliffs and frost-shattered pebbles on beaches and in river beds.

A central characteristic of the periglacial zone is the existence of permanently frozen ground, or *permafrost*. The definition of permafrost is based purely on sub-zero ground temperatures. It becomes geomorphologically significant only when the frozen ground is 'wet', in other words when bedrock, regolith and soil voids are ice-filled. Such ice transforms a porous rock such as chalk into an impervious mass and also solidly cements together mixtures of minerals and organic matter (*i.e.* soils and regolith). So solid is it that Baird dryly wrote, 'The explorer who succumbs to the severity of the climate must be buried above ground or with the benefit of dynamite'. The significance of permafrost lies firstly in its extent, it occurs under 26% of the earth's surface, and secondly in the influence it has on processes which take place above it.

During the summer the upper layers of the ground slowly warm from the surface down. This zone of seasonal thawing, *the active layer*, varying in depth from a few centimetres to several metres, is the location of a range of processes which are described below.

The most widespread process is that of *solifluction*. As the summer thaw begins the upper part of the soil

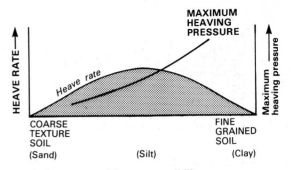

6.32   Soil texture and frost susceptibility

becomes saturated with water unable to drain away because of the still frozen ground beneath. Tills with a high proportion of silt and clay can become so waterlogged that they develop a porridge-like consistency and begin to flow on quite gentle slopes. The movement is quite slow, about a centimetre a year on 2° slopes and 10 cm on 10° slopes. This shallow flowage often occurs as a series of lobes. A critical factor is moisture supply. Meltwaters downslope from a lingering snowpatch, for example, will allow solifluction throughout the summer whereas adjacent slopes have partially dried out and solifluction slowed as the thawed zone deepens during the summer. Solifluction is one of the processes occurring around and beneath the snowpatch which can cause its 'countersinking' into the hillside as a nivation patch. Although slow-moving and occurring only for short periods in the summer, solifluction has an important smoothing effect on relief as material is sludged downslope.

With the onset of winter the surface of the active layer freezes and the cold gradually penetrates to the permanent 'frost table'. Since ice occupies a greater volume than water the freezing of the interstitial water causes the ground surface to be *heaved* upwards. Just how much heaving there will be will depend on the grain sizes of the material involved as can be seen from inspecting Fig. 6.32. The engineering implications of this susceptibility will be explored in the final chapter.

As the winter cold gradually penetrates the ground a layer of water-saturated material may become trapped between the advancing frost and the permafrost beneath. It may be under such intense hydrostatic pressure that it fractures the overlying layer and bursts through onto the surface where, in the intense winter cold, it rapidly freezes as a '*surface icing*'.

Roads built without a sensitive appreciation of active-layer dynamics may be rendered impassable by such icings up to 1 km long and 3 km thick.

In some situations where water migrates laterally through silts, distinct *ice lenses* may grow. These can produce localised bulging of the surface. If we consider a mixture of stones and silt the penetration of the winter freezing front will be uneven. The release of latent heat connected with the freezing of water-saturated silts will delay freezing so that water beneath stones will begin freezing sooner. Water from the surrounding silts will then migrate to the ice crystals under the stone and begin to form an ice lens. Its growth can push the stone upwards towards the surface, at a rate of up to 5 cm a year. This process sorts material and contributes to the growth of patterned ground which will be mentioned later. Ice lenses may form in the soil for other reasons reflecting the penetration of cold and movement of water, such as differences in snow and vegetation cover.

The growth of ice lenses over a period of years leads to the formation of a whole family of landforms. At the smaller end of the spectrum, up to 50 cm high, are the *tundra hummocks*. These swarms of ice-cored tussocks are covered with cotton grass and sedge. As the hummock grows its top may become desiccated, the vegetation dies and with its insulative effect gone the ice-core thaws and the hummock degrades—a cycle probably taking about 60 years.

The largest ice-cored features are the *pingos* but as they require a special range of conditions they are not

widespread. The largest collection, about 1,400, exist in the Mackenzie Delta on Canada's arctic coast.

The development of these ice-cored elliptical mounds ranging in size to 70 m in height and 700 m across is summarised in Fig. 6.33. The sequence begins with the gradual draining of a shallow lake, permafrost advances from the sides as it shallows and its insulative effect on the ground beneath is reduced. This encroachment gradually creates pressure so that the water is forced upwards as a kind of 'hydrolaccolith'. The ice blister is covered with silt, peat and vegetation. With continued growth of the ice lens, (upwards at a rate as high as 50 cm a year), the overlying layers are ruptured. Deprived of its summer insulation the ice-core may begin to thaw and the pingo degrade.

The suggestion has been made above that ice-cored features may have a cyclical development, ice growth altering the vegetation so that insulation against summer warmth is reduced leading to the decay of the feature. Vegetation disturbance and climatic change may also cause other types of ice lenses within the soil to melt. Such lenses can be up to 10 m thick and hundreds of metres in horizontal extent. When they gradually thaw the ground settles and hollows or closed depressions are formed. Such terrain is known as *thermokarst*.

As ground ice melts an irregular shallow lake may be formed. Summertime wave action soon smooths its shores and the *thaw lake* can grow up to 2 km or so in diameter, although they usually remain less than four metres deep. As time passes the lake may be filled

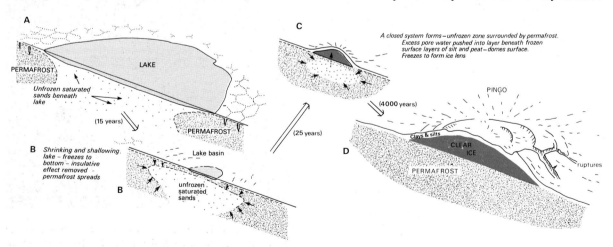

6.33   The growth of a closed system Mackenzie pingo

by vegetation or it may be drained and vegetated as it becomes incorporated into the area's stream network. In both cases the outcome is that permafrost begins to re-establish itself. The birth, growth and death of thaw lakes is in fact one of the most active cyclical processes in operation on the flatter tundra.

Mention was made earlier of the sorting process at work in the active layer with the upfreezing of stones and the migration of fines ahead of the freezing front. This is one of the range of processes leading to the formation of *patterned ground*. This term is applied to the sorted and unsorted circles, nets, polygons and stripes which are common features of the periglacial environment. They are complex in origin and different processes may produce the same shape, as is suggested in Fig. 6.34. On level ground circles, typically ranging

from 0·5 to 3 m across, may be found. These may be sorted, with stones at their borders, or unsorted, when they might be marked by surface doming and a vegetated border. Coalescing circles may produce a network pattern and on slopes the pattern is transformed into downslope striping.

In extremely cold areas, *i.e.* with winters down to −40°C, intense thermal contraction of the frozen ground may produce a series of cracks similar to those formed in drying and shrinking clay. In the summer these cracks may fill with water which freezes at depth to form an *ice wedge*. Successive winters cause further contraction and the crack may deepen at a rate of 1–20 mm/yr, eventually extending to a possible depth of 30 m. The wedges widen too, to such an extent that

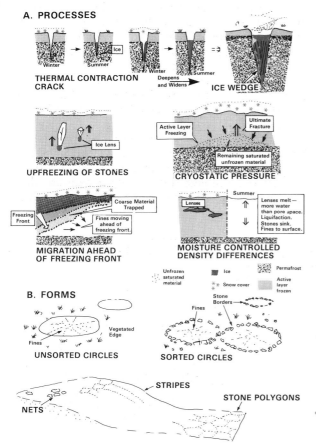

6.34   Patterned ground

6.35   Tundra polygons revealed by snow cover differences on the flood plain of the Colville River, Alaska. This vertical view also shows dramatic signs of river channel migration

in some areas the ice volume in the wedges exceeds the intervening soil volume. Viewed from the air these wedges form an irregular hexagonal pattern, the *tundra polygons*. Up to 40 m across they are usually 'high edged' and lower in the centre. When climate changes thawing of the wedge may transform them into high-centred features. A complete thaw will fill the wedges with sands and gravels washed into the crack. Finding such filled wedges is particularly useful when attempting to reconstruct past environments since tundra polygons today are found mainly in extremely cold areas such as the north slope of Alaska (Fig. 6.35).

Other agencies at work in the periglacial environment

Low temperatures, limited organic life and the lack of circulating ground waters are three facets of the periglacial environment which are reflected in the character of *tundra soils*. These are frequently falsely placed in the zonal soil order although by definition zonal soils can only form under free-drainage conditions. Impeded drainage is in fact the norm, affecting about 90% of the tundra area so gleys and bog soils occupy most of the area.

The emphasis in the previous section was on the effects of ground ice but it is important for a balanced view of the periglacial world that the other processes be considered. *Chemical erosion*, with low temperatures, limited organic matter production and restricted water circulation, is frequently overshadowed by mechanical processes. The rates of silica removal from granite shield areas, for example, are equivalent to a rate of surface lowering of 0·8 mm/1,000 years out of a total removal rate of 36 mm/1,000 years. Limestones are a little different since cold water absorbing more CO_2 is potentially more aggressive. However, the solutional effects visible in karst landforms are subdued by the shortness of the liquid water season and the recency of exposure after glaciation.

Restricted vegetation cover, broad valleys floored with stream and solifluction material, together with sandy outwash plains means that *wind* can easily entrain material. Strong katabatic winds blowing out from the former ice sheets could pick up and carry sand grains which can abrade rock to produce flutings and grooves. Small erosional features might therefore be produced close to the surface. Sandy outwash material can also be reworked by the wind to produce dunes. The finer grain sizes, on the other hand, may be transported over greater distances and such loess may be deposited at some distance from the periglacial zone. Largely bordered today by the Arctic Ocean, and not an ice sheet, the supply of sediments susceptible to wind action in the present periglacial zone probably bears no relation to that available in the past.

Finally, but far from last in terms of significance, is the action of *running water*. In arctic areas today about 80% of the year's runoff occurs in only 3% of the time. Such peak discharges following snowmelt can be very effective in both erosion and transport. Flowing over tills, outwash and solifluction deposits the streams are liberally provided with rock material. As discharge falls during the autumn the streams' velocities and competence are reduced and the load is dumped in and around the shrinking stream where it can be acted on by frost-shattering and the wind. An impermeable barrier of permafrost beneath the surface could also speed runoff responses from the catchments. This further accentuates the peaky hydrographs of the streams.

The extent of the periglacial zone

A useful way to delimit the periglacial zone is to map

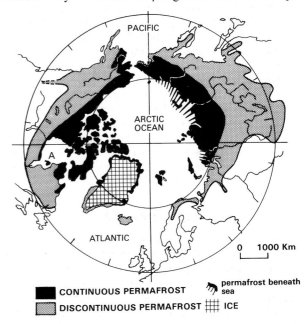

6.36 Permafrost distribution

the *present distribution of permafrost*. Figure 6.36 indicates the extent of both continuous and discontinuous permafrost. The latter is thinner and occurs as patches intermingled with unfrozen ground.

What controls this distribution? *Climate* is a basic factor, but is it simply a matter of air temperatures? Permafrost could be encouraged either by low winter temperatures or by cool summers. Snowfall, too, can be significant. Look at Fig. 6.36 and compare it to mean annual isotherms. There is a northward displacement of permafrost on the east side of Hudson Bay (point A). This reflects the insulative role of heavy early winter snowfalls over Ungava before the Bay freezes over. In other situations heavy winter snowfalls could actually promote permafrost by hindering summer warming. In other words the various climatic parameters may substitute for one another!

At smaller scales *topography* and *vegetation* are important influences on distribution. Sites of different elevations and of different exposures will have differences in solar radiation and snow cover. Vegetation, itself a function of climate, topography, rock and time, has a crucial role. The lichen mat which covers extensive areas of the arctic and subarctic acts almost like a light green foam-rubber mattress: in summer when dry it insulates the ground against summer warming, in winter when saturated and frozen it allows the conduction of winter cold into the ground. It is for these reasons that preservation of ground cover is so important in permafrost engineering if the deepening of the active layer is to be avoided.

To add to the complexity there is the problem of *time*. It takes 10,000 years for permafrost 500 m thick to form and in northern Siberia the permafrost is three times this thickness. Figure 6.1 indicated vast areas of ice-free surface in Siberia at the height of the last glaciation. The existence of permafrost 800 km south of the present tree line at 45°N may therefore reflect the climatic history of the last 20,000 years.

The legacy of past permafrost

At the height of the last glaciation areas around the ice sheets were affected by the processes described earlier. As climate changed permafrost melted, river regimes altered and vegetation belts migrated. The recency of this transformation means that many details of the mid-latitude environments are legacies from this periglacial past.

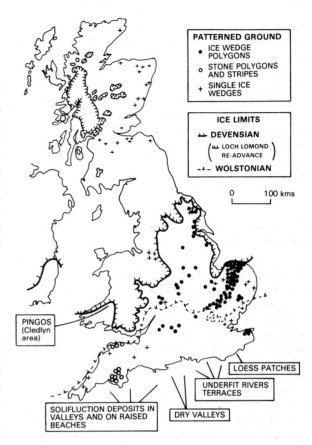

6.37   Some of the periglacial legacy in Britain

Figure 6.37 shows the distribution of some of these relict features in Britain. The identification of fossil tundra polygons has relied heavily on aerial photography. The infilled wedges have different moisture characteristics which may be reflected in vegetation differences although the actual ground configuration shows no signs of the polygons.

In addition to the signs of frost action and the previous existence of permafrost the changes in river behaviour are equally striking. As bedrock thawed, limestones and chalk, previously rendered impermeable by the permafrost, became unable to support as many surface streams and *dry valleys* were formed. The regimes of rivers changed too as runoff was no longer concentrated in the short melt seasons. The thawing of the permafrost and the thickening of the vegetation cover also altered the various stores and transfers within the basin hydrological cycles. The reduction in the channel forming discharges brought with it a host of channel changes. One result is the existence of *'underfit'* rivers which meander within much larger amplitude valley meanders.

Changes in runoff behaviour were accompanied by a reduction in the load supplied to and carried by the streams. Streams in many cases began cutting into the deposits which their periglacial ancestors had left flooring valleys, producing *terraces* on either side of their current flood plains. As late glacial climate fluctuated with minor readvances of ice so too did the rivers' behaviour. When the fact of multiple glaciations and interglacials is added to the existence of sea-level fluctuations the combinations of possible aggradation and incision becomes very complex, especially in the case of larger river systems. Untangling the terrace sequence frequently demands the kind of detective work hinted at in the discussion of Scottish raised beaches. For a fuller discussion of this topic the further readings can be consulted.

Periglacial mass masting, particularly solifluction, has also left its legacy in the form of valley fillings like the *'combe'* plugging the bottom and sides of many chalk valleys. *'Head'* deposits flooring valleys, slumped over beaches and fanned out where valleys debouch onto lowlands are also a periglacial legacy.

Solifluction may also have left its imprint in the form of *valley asymmetry*. The depth of the active layer and ground moisture, both of which are influenced by exposure and aspect, may well vary on opposite sides of a valley with inevitable consequences for the solifluction process. Figure 6.38 models the situation in northern Canada. If the annotation is studied you will realise that the situation is quite complex. Interpreting valley asymmetry on the north slope of the Chilterns may therefore be no easy matter.

THE QUESTION OF ARIDITY: PAST AND PRESENT PROCESSES IN THE SAHARA

Apart from a coastal excursion this chapter has focused on glacial and periglacial environments. Before leaving the Pleistocene, however, one major question remains—what type of environmental changes affected the tropics? Facets of this problem will now be examined in the context of the Sahara.

The central parts of the Sahara desert today are hyper-arid and they are fringed by extensive semi-arid zones. The term *arid* is a relative one, reflecting the partial breakdown in the hydrological cycle so that water losses exceed inputs and there is no perennial drainage network. Precipitation in the Sahara is characterised by low gross totals, with annual means of less than 50 mm over the arid zone itself. Equally important is the infrequency and irregularity of precipitation. The word desert implies limited life. When the scanty vegetation cover is combined with the character of the area's rainfall, surprisingly intense (if shortlived) surface runoff can occur.

Meteorologically the Sahara is also characterised by low cloud cover and high insolation. This leads to harsh diurnal temperature ranges, of over 55°C, and dramatic humidity changes from as low as 5% during the day to over 60% at night. The climatological cause of aridity, in the case of a tropical desert like the Sahara, lies in the general circulation of the atmosphere. The arid zones occur beneath the subsiding air of the subtropical high-pressure cells. The trade winds associated with the cells are not particularly strong, the geomorphological significance of wind stemming (as with water) from the exposed surfaces and paucity of vegetation.

Finally, in any discussion of the Sahara it is important to remember its sheer size and diversity. Its landscapes range from volcanic areas like the Hoggar and Tibesti massifs to below-sea-level depressions like the Qattara. The dominant topographic element, however, is plain, whether the hamada (stony flat

SUN

Aspect	a) High Latitude		b) Sub-Arctic	
Aspect	SOUTH FACING	NORTH FACING	SOUTH FACING	NORTH FACING
Sun	more	less	a lot	less
Active Layer	deeper	shallow	no permafrost	present
Solifluction	more	less	none	exist
SLOPE	GENTLE	STEEP	STEEP	GENTLE

SNOW
(Wind from W.)

	(i) Early Thaw		(ii) Delayed Thaw	
Aspect	EAST FACING	WEST FACING	EAST FACING	WEST FACING
Accumulation	more	less	more	less
Thaw	longer	shorter	limited	more
Solifluction	last longer	restricted by lack of moisture	less	more
SLOPE	GENTLE	STEEP	GENTLE	STEEP

6.38 The complexities of valley asymmetry (see also Fig. 3.34)

plateaus), reg (pebble plains) or erg (sand).

It is reasonable to assume that the high-pressure cells have been an enduring feature of the earth's atmosphere. The question is has their intensity and position varied—in the longer term as plates and continents moved, and in the shorter term as climate oscillated in the Pleistocene? The existence of *pluvials*, wetter (and/or cooler) periods during the Pleistocene, has long been recognised. How are these pluvials related to events in higher latitudes? In the case of the central Sahara do they represent an extension southwards of the 'Mediterranean' winter rainfalls or an extension northwards of the ITCZ and savanna summer rainfalls? The former might relate to glacial conditions further north, the latter to an interglacial. This question is in fact quite complex and will not be explored further here. The focus of the remaining section is solely on the evidence and significance of the pluvials, which is summarised in Fig. 6.39.

A fundamental principle in reconstructing past environments is to have an understanding of the present. Assessment of the current streamflow and *runoff* situation is limited by two gaps in our information. The first concerns precipitation data: with up to 1,000 km between rain gauges detailed knowledge of the nature and extent of Saharan rainfall is impossible. The second concerns the way in which the desert surface responds to the rain falling on it.

It is a well worn truism that the study of deserts is bedevilled by mirages. One of these concerns misconceptions about 'torrential' desert rainfalls: summer thunderstorms in the Hoggar have been known to drop 5 mm of rain in ten minutes and Aouzou in the Tibesti recorded 370 mm in three days in 1934. In

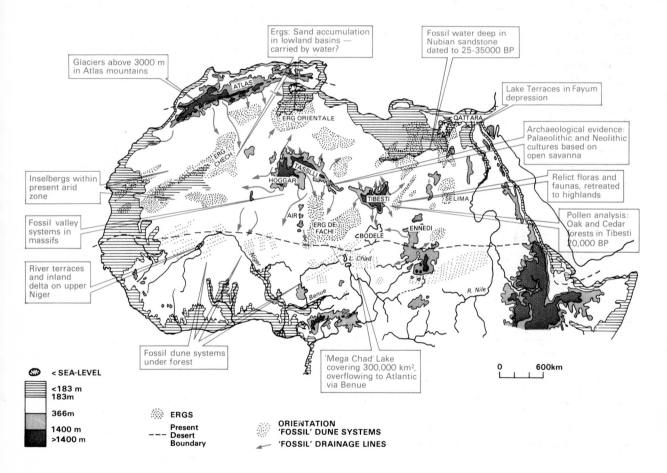

6.39 The evidence for Saharan climatic change

reality, however, most of Algeria has rainfall intensities less than those recorded in France. In fact there may be many light showers which fail to reach the 'runoff threshold' and it has been suggested that only about 10% of desert rainfall generates surface runoff. Yet there is ample evidence for the work of running water. How can this apparent paradox be resolved?

One factor in the response of the desert surface to rainfall is the high proportion of exposed bedrock, reflecting sparse vegetation and removal of the finer weathered materials by the wind. Such rocky slopes can shed water freely, but the response of the surfaces littered with weathered material (sand) is a little less obvious. Humid temperate ideas of the importance of antecedent moisture and high infiltration rates on sandy textured material are unhelpful. Israeli work in the Negev indicates that 'soils' on flatter ground may be surprisingly compact and infiltration rates of 2 mm/hr have been recorded! The lack of vegetation roots and the mixing activities of soil fauna play their part in this, but the effect of rain beat in tamping down the surface and blocking pores is more important. The discovery of such low infiltration rates is crucial, since it helps explain how low-angle slopes of less than 2° can generate more runoff than adjacent 20° slopes. Secondly it helps us understand how we can get flooding in deserts from rainfall events which are non-exceptional in terms of their intensities or durations. In summary, therefore, desert surface-water flow is immediate and because of the rainfall pattern rather shortlived. The amount of deep ground water recharge in environments where potential evapotranspiration is measured in metres and rainfall in millimetres is negligible. Interestingly water deep in the Nubian sandstone has been dated (by radioisotope methods) to 25–35,000 years BP, suggesting it is 'fossil' water dating from a Pleistocene pluvial.

The immediacy of runoff after the threshold has been exceeded takes a number of forms. One of these is *dispersed wash* where a multitude of intertwining ribbons of water flow over the surface. This shallow film affecting the whole surface can entrain fine material, although its actual erosional effect is very limited. Downslope, however, the increasing volumes of water become concentrated into rills excavating channels up to a metre in depth. The individual rills may be used only once so that rill action from storm to storm can eventually affect the whole slope.

Another runoff response is the well documented *stream flood*. In hilly areas the water rushes down the valleys (wadis) and the frontal wave of the advancing flood quickly picks up rock waste mantling the wadi floor. Sediment concentrations can be extremely high, figures of up to 50% have been recorded—virtually debris flow! As this concentration reduces the weight of submerged boulders by up to 60% quite large rocks can be carried by the stream flood. Once the stream emerges from the confining valley walls its work is mainly depositional in the form of fans, cones and spreads of material on the lower ground. At the present time such stream floods can do a lot of work quickly but between these periods of life there may be years of 'suspended animation' when the stream system is dormant. Between the storm events there is limited weathering on the wadi side walls which are in consequence rather gorge-like, although this varies with local factors like lithology and structure.

Current water action of this type is well documented but more central to the theme of the chapter there is *evidence for wetter periods in the past*. In the case of the Tibesti and Hoggar massifs, for example, there are extensive valley systems far too big to have been carved and integrated by the present episodic rainfalls. As Fig. 6.39 shows these former water courses radiate outwards from the mountain areas.

In the central Sahara such lines can be followed to a series of depressions. Streams from the Hoggar, Tibesti and Ennedi appear to have flowed towards the Bodele and Chad depressions. As Fig. 6.39 indicates there is evidence for the existence of a 'Mega Chad' lake covering over 300,000 km² with shorelines more than 50 m above the present shrunken remnant.

In addition to the range of erosional and depositional evidence for the previous existence of river systems and lakes there is also some 'biotic' evidence for wetter and/or cooler conditions. In the central Sahara a belt of former lakes and swamps existed between the Tassili plateau and the volcanic Hoggar massif. Archaeological evidence suggests that the area was occupied by palaeolithic hunters and the remains of their prey (antelope, gazelles, rhinos and elephants) suggests Kenyan-type mammalian fauna. Later neolithic sites also suggest a savanna environment and petroglyphs (carvings scratched into the desert varnish of the rock shelters) depict hunting scenes set in an open savanna biome. Pollen analysis from former lake sites in the Tibesti also indicates the existence of oak and cedar forests dating to about 20,000 BP. Finally, there is the existence of relict faunas and floras from this more humid past. The small crocodiles of the Tibesti and Ennedi could hardly have trekked there across the desert, the reptiles became trapped on their slightly more humid islands as the desert 'sea' expanded at the end of the most recent pluvial. The detailed evidence in fact suggests several pluvial periods, the last major one ending about 15,000 BP during which time the central Sahara supported open savanna on the higher and wetter sites.

Desert *weathering* is a less-fruitful field of evidence for climatic change since the present rate of operation of the processes is unclear. Water is never entirely absent today and forms of rock weathering like block disintegration and exfoliation are recognised as being water assisted. Many desert rocks are however covered with a dark red-black layer of *varnish*, a thin film of iron and manganese oxides and silica formed by the evaporation of mineralised water migrating from within the rock. As the water evaporates salts are left on the surface, the soft halides are removed by the wind leaving the more durable iron, manganese and silica to be polished by wind action. Some authorities believe the alternating wet and dry savanna climate provides the best environment for the creation of this desert varnish. Neolithic 'graffiti', as fresh as the day it was etched into the varnish, might confirm this view. However, the role of moisture from dew (which can exceed annual rainfall as a moisture source) complicates the issue. At the largest scale landforms such as the *inselbergs* (isolated residual hills), occurring within the present desert area, have been considered indicators of past pluvial conditions. Moisture-assisted weathering at the base of the inselbergs in a savanna environment is believed to be significant in their formation. Assigning inselbergs to the list of 'relict' features is a complex issue which can be followed up in the further reading.

The final area concerns the role of *wind*, a field absolutely shimmering with mirages. One misconception concerns the proportion of sand areas in deserts: the ergs of the Sahara occupy less than a fifth of the area. One problem is where has this sand come from? Two primary sources for the quartz grains are granite and the disintegration of sandstones. The problem with the former is that desert weathering is weak once the particle size drops below 10 mm so the second source is most significant. In the case of the Sahara this is the Nubian Sandstone.

Given the theme of this section a more relevant dimension to the question of sand origin is the actual location of the ergs. How has the sand been carried to them? The ergs occupy lowland basins and sand accumulation in them has demanded centripetal accumulation which wind is unable to do. As Fig. 6.39 suggests, water would have been the only agency capable of concentrating the sand in these sites.

What therefore is the present role of wind? At the smaller scales sand *abrasion* polishes, pits and smooths exposed rock in the zone close to the ground, where grains are being carried by saltation and surface creep. At the medium scale its role has been overstressed, moisture assists cavity weathering and the wind removes the weathered debris. Some soft lacustrine sediments, however, do appear to have been carved into elongated grooved furrows or yardangs. At the larger scale its prime role is *deflation*, the physical removal of finer debris. Pebble-strewn reg surfaces have been attributed to wind removal of fines but they could result from runoff or sorting processes—another example of equifinality of form. The more dramatic deflation features, such as the Qattara Depression whose formation has involved the removal of 3000 km^3 of material, can be more positively linked to wind action.

An important current role is *dune formation*. No evidence exists that the ergs are expanding so the wind is probably only moving and reshaping the sand. The pioneer work on the dynamics of this shaping (by Bagnold) inadvertently set the tone for later descriptions which have overstressed two dune forms, the elongated sief and the crescentic barchan. The latter according to Warren occupy a mere 0·01% of the desert sand area! Dunes are best viewed as equilibrium bedforms developing in a system comprising of loose sand in a flowing medium—the air. They show a regularity with three universal elements, ripples, dunes and draas (Fig. 6.40) and in plan view crescents opening downwind (barchanoids) and upwind (lingoids).

A question relevant to the theme of this section is can we match dune patterns with wind direction? *Aklé* is the name applied to dune networks in the western Sahara. Figure 6.40 shows how this results from the interaction of two components of the wind, namely eddies transverse to the wind direction and vortices of convergent and divergent flow parallel to its flow. Convergent (slow) flow produces barchanoid elements and divergent (fast) flow the more subdued lingoid elements. The lower part of the diagram shows that two axes may appear in the pattern to produce elongated dune lines. If one of the axes of the dune crests becomes emphasised a parallel system of relatively straight dune ridges, or *siefs* is produced. Earlier views of these longitudinal elements linked them with the resultant of the annual wind pattern but the existence of two sief lines intersecting at acute angles (as in Libya) suggests the relationships suggested in Fig. 6.40 operate.

Figure 6.39 showed the existence of *fossil dunes* now covered by vegetation and undergoing chemical weathering. These indicate a previous southerly extension of the Sahara. The orientation of these dunes in a series of vast arcuate systems holds clues for the past wind circulation. As information on present wind directions and frequencies is limited reconstructing past climates on this basis is far from easy.

Element	Spacing (wavelength) (m)	Height	Names
transverse longitudinal	0.01 - 3	0.00001 - 0.2	RIPPLES
transverse longitudinal	20 - 300	1 - 30	DUNES
transverse longitudinal	1000 - 3000	20 - 200	DRAAS

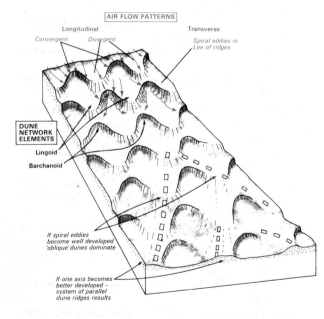

6.40 Bedform networks

CONCLUSION

This chapter has ranged widely in scale and type of landscapes discussed, from the world of snow and ice to the subtropics and from small ice wedges to the immensity of the Sahara. Its *unifying theme* however has been the magnitude and significance of the world-wide *climatic changes* of the Pleistocene. From time to time comment has also been made about the nature of *evidence* and the processes of *explanation*. The difficulties of this have been rather glossed over. How can the present be the key to the past when we have no current equivalent to the Wisconsin ice sheet to investigate and when physiographic evolution may have slowed so much in deserts that measurement of microscopic changes is required? Human ingenuity has and will continue to grapple with this problem—after all, understanding climatic change may well be vital to mankind's survival.

Review Questions

1. Assess the role and effects of meltwater in the glacial environment.
2. With reference to Fig. 6.7, *a*) describe the present glacier features, *b*) outline a possible sequence of ice decay, and *c*) describe the landforms created and exposed during this deglaciation.
3. What techniques are available to help in constructing a deglaciation sequence?
4. Describe how permafrost influences the processes at work in the periglacial landscape.
5. Describe and account for the pattern of ice motion in cirque and valley glaciers. What are the erosional and depositional consequences of this movement?
6. 'The significance of continental ice sheets does not lie in their erosional activities.' Discuss.
7. Review the evidence for the existence of meltwater lakes.
8. Describe the pattern of glacial isostatic movements. What evidence do we have for these?
9. Assess the significance of past and present water action on the shaping of desert landscapes.

Research Questions

1. Cirque distribution may reflect *a*) regional snowlines, *b*) aspect and exposure effects on snow accumulation, and *c*) lithology. Try and formulate some ideas reflecting these (*i.e.* snowlines will be lower towards the north and towards the moisture

source; more snow could accumulate on lee slopes, *etc.*). Then design a map analysis programme to test your ideas, perhaps along the lines of Fig. 6.41.

2. *a)* Using the data in Fig. 6.42 plot isolines to show the decay pattern of the Scandinavian ice (*i.e.* lines to show ice margin 10,000, 12,000 BP, *etc.*). Comment on the uplift rates and erratic traces, what do they suggest about ice thickness and flow direction?

   b) What have been the erosional and depositional

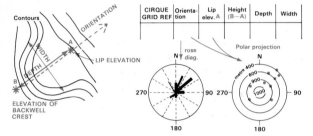

6.41 Cirque analysis

6.42 The Scandinavian ice sheet

landforms produced beneath and at the margins of this ice sheet?

3. Produce a map to show the extent of Pleistocene ice in Great Britain. What landform legacy has it left in both highland and lowland Britain?

4. With reference to a specific coastline outline the effects of Pleistocene sea-level changes.

5. Describe the origin and form of coral reefs and atolls.

6. Using the ideas summarised in Fig. 6.38 devise a field programme to investigate valley asymmetry.

7. Describe the sequence of post-glacial climatic changes and their effects on the distribution and character of British vegetation.

Further Reading

Andrews, J. T., 1975, *Glacial Systems*, Duxbury–Wadsworth.

Baird, P. D., 1964, *The Polar World*, Longman.

Barry, R., & Ives, J. D. (Eds.), 1974, *Arctic and Alpine Environments*, Methuen.

Bird, J. B., 1972, *The Natural Landscapes of Canada*, Wiley.

Brunsden, D., & Doornkamp, J. (Eds.), 1973, *The Unquiet Landscape*, David & Charles.

Cooke, R. U., & Warren, A., 1973, *Geomorphology in Deserts*, Batsford.

Embleton, C. E., & King, C. A. M., 1975, *Periglacial Geomorphology*, Arnold.

Flint, R. F., 1971, *Glacial and Quaternary Geology*, Wiley.

Goudie, A., & Wilkinson, J., 1977, *The Warm Desert Environment*, Cambridge University Press.

King, C. A. M., 1972, *Beaches and Coasts*, Arnold.

Lewis, C. A. (Ed.), 1970, *The Glaciations of Wales*, Longman.

Price, R. J., 1972, *Glacial and Fluvioglacial Landforms*, Oliver & Boyd.

Sissons, J. B., 1976, *Scotland*, Methuen (plus other titles in the series *Geomorphology of the British Isles*, Ed. K. Clayton).

Sparks, B. W., & West, R. G., 1972, *The Ice Age in Britain*, Methuen.

Washburn, A. L., 1973, *Periglacial Processes and Environments*, Arnold.

West, R. G., 1977, *Pleistocene Geology and Biology*, Longman.

Sugden, D., & John, B., 1976, *Glaciers and Landscape*, Arnold.

7 Man, Technology and Environment

Some of the ways in which man has impinged on and been influenced by environmental systems forms the focus of this chapter. The relationship between man and environment, referred to in many places earlier in the book, becomes the central theme here. This will be explored by examining three problems—permafrost engineering, atmospheric modification and marine flooding. It is, however, the general principles underlying these which are of major importance.

THE PROBLEMS OF PERMAFROST ENGINEERING

With 47% of their country underlain by permafrost the Russians acquired an early lead in understanding the character and distribution of frozen ground. The Canadian and Alaskan northlands remained relatively unknown environments until strategic considerations and resource exploitation during the last three decades produced a rapid growth in their populations.

Continuous permafrost is found where mean temperatures lie below $-8°C$ (Fig. 6.36). Discontinuous permafrost is encountered in 'warmer' areas wherever the surface radiation balance produces a ground temperature below $0°C$. Its geomorphological role was discussed earlier, its engineering significance forms our theme here. In the discontinuous permafrost zone, where thawed islands (or taliks) exist *site factors* are especially important in controlling the distribution of ground ice. They include those which reflect ground insulation (snow cover, vegetation, soil moisture and lakes) and those reflecting solar energy receipts (slope aspects, surface albedos, etc.).

Placing even a small unheated building on permafrost can be problematic as it changes the radiation balance of the surface. Summer thawing under the building is slowed, especially on the shaded side, and the frost table grows upwards. The weight of the building on a saturated active layer during summer may therefore cause slumping and tilting. With a heated structure this problem is more severe as heat escaping into the ground actually lowers the frost table. Also during late winter, when the active layer has frozen around the building, an unfrozen area exists beneath it which can bodily heave the structure when it finally freezes. At best doors are difficult to open and floors slope, at worst the whole building may collapse. Ruptures (icings), when water-saturated material bursts through the thinnest part of the downfreezing layer, can occur beneath the building. Breaking through the floor it can cause considerable disruption!

Roads are equally prone to damage and maintenance costs are high. The 3,000 km Alaska Highway linking Fairbanks with the Canadian road network at Dawson Creek was built under the pressure of wartime priorities between March and November 1942. Detailed site surveys were impossible and since even the term permafrost had yet to be coined in the West it is not surprising its route failed to reflect the effects of ground ice. Such disregard of environmental factors have proved expensive, maintenance swallowing $150,000,000 during the first two decades of the Highway's existence.

To reduce curves and gradients roads use combinations of cuts and fills. Fills (embankments) were susceptible to two processes as the permafrost rose into the embankment. The bulge in the frost table beneath the road causing summer flooding on the upslope side. On the other side drier conditions encouraged thawing and the roadbed disintegrated as it slumped downslope. With cuts the autumn downfreezing meant an early fusing of the seasonally frozen ground with the permafrost beneath at the 'heel' of the cut. Water moving downslope between the advancing winter frost and the permafrost would break through above this 'pinching off' causing an icing across the road. To add to these changes in the dynamics of the active layer there is the problem of frost heaving and the growth of ice lenses in the road bed. As Fig. 6.32 showed some soil textures are particularly susceptible to these processes and the consequent decay of the road.

These kind of problems and their solutions are shown in Fig. 7.1. Permafrost engineering can follow four courses, the first being to ignore the ice, with the kind of consequences summarised in Fig. 7.1! Elimination, by steam jets thawing the ground or by stripping vegetation to encourage natural thawing, is only possible with small patches of 'relict' permafrost which will not reform. *Passive* approaches on the other hand try to preserve the permafrost and design special structures. Preservation means insulating permafrost from the thermal consequences of construction and it can take a number of forms. For smaller buildings a well-drained gravel pad 0·5–1·5 m thick may suffice, aided by such stratagems as painting the structure white and aligning its long axis N–S to minimise its effects on the ground's radiation balance. Alternatively steam jets can 'drill' holes into the permafrost. Piles inserted in these are allowed to freeze into the permafrost and buildings, or water and sewage lines, can be erected on top of them. The space

7.2 Permafrost construction, Inuvik, Canada. The igloo shaped church is on a gravel pad, the white utildor (containing insulated water and sewage pipes) and the building at the right are raised on piles

beneath insulates the ground from the effects of the building's heat and the permafrost is preserved.

Roads on the other hand can hardly avoid running across the surface and their deterioration can be slowed by the techniques shown in Fig. 7.1. Airstrips present similar problems and are best constructed of gravel (0·5 to 3 m thick) laid over undisturbed ground. The gravel should preferably have a high albedo to reduce surface temperatures in summertime.

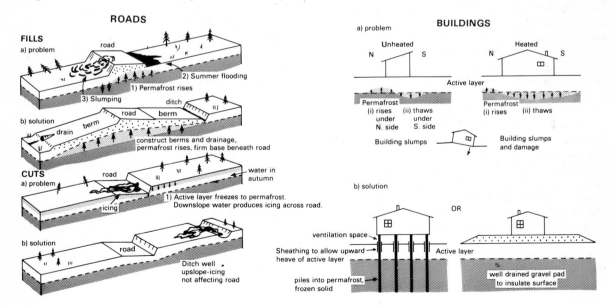

7.1   Construction problems and solutions in permafrost regions

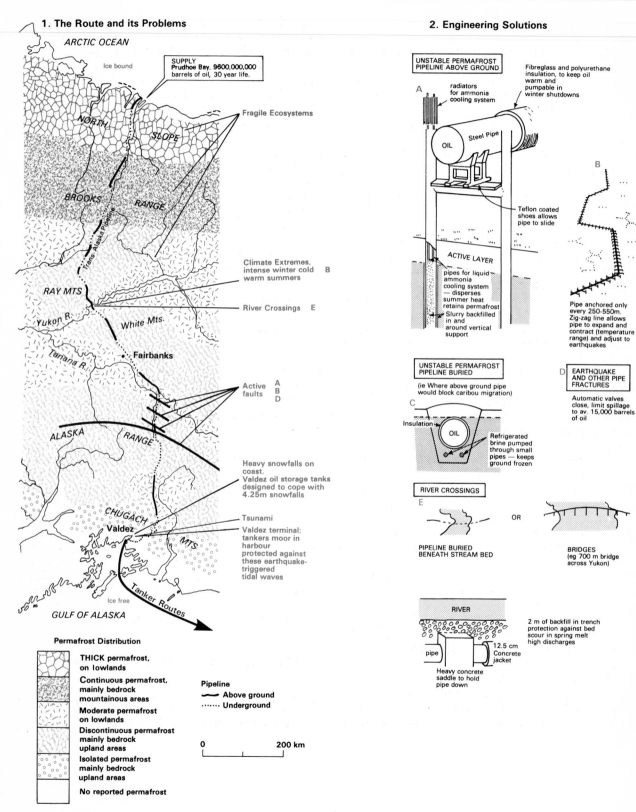

1. The Route and its Problems

ARCTIC OCEAN

Ice bound

SUPPLY
Prudhoe Bay, 9600,000,000
barrels of oil, 30 year life.

NORTH SLOPE

Fragile Ecosystems

BROOKS RANGE

Trans Alaska Pipeline

Climate Extremes,
intense winter cold
warm summers   B

RAY MTS.

River Crossings   E

Yukon R.   White Mts.

Tanana R.

• Fairbanks

Active   A
faults   B
           D

ALASKA RANGE

Heavy snowfalls on
coast.
Valdez oil storage tanks
designed to cope with
4.25m snowfalls

CHUGACH
Valdez   MTS.

Tsunami

Valdez terminal:
tankers moor in
harbour
protected against
these earthquake-
triggered
tidal waves

Ice free

Tanker Routes

GULF OF ALASKA

Permafrost Distribution

THICK permafrost,
on lowlands

Continuous permafrost,
mainly bedrock
mountainous areas

Moderate permafrost
on lowlands

Discontinuous permafrost
mainly bedrock
upland areas

Isolated permafrost
mainly bedrock
upland areas

No reported permafrost

Pipeline
—— Above ground
······· Underground

0   200 km

2. Engineering Solutions

UNSTABLE PERMAFROST
PIPELINE ABOVE GROUND

Fibreglass and polyurethane
insulation, to keep oil
warm and
pumpable in
winter shutdowns

A   radiators
     for ammonia
     cooling system

Steel Pipe

OIL

B

Teflon coated
shoes allows
pipe to slide

ACTIVE LAYER

pipes for liquid
ammonia
cooling system
— disperses
summer heat
retains permafrost

Slurry backfilled
in and
around vertical
support

Pipe anchored only
every 250-550m.
Zig-zag line allows
pipe to expand and
contract (temperature
range) and adjust to
earthquakes

UNSTABLE PERMAFROST
PIPELINE BURIED

(ie Where above ground pipe
would block caribou migration)

C

Insulation

OIL

Refrigerated
brine pumped
through small
pipes — keeps
ground frozen

D   EARTHQUAKE
    AND OTHER PIPE
    FRACTURES

Automatic valves
close, limit spillage
to av. 15,000 barrels
of oil

RIVER CROSSINGS

E

OR

PIPELINE BURIED
BENEATH STREAM BED

BRIDGES
(eg 700 m bridge
across Yukon)

RIVER

2 m of backfill in trench
protection against bed
scour in spring melt
high discharges

pipe

12.5 cm
Concrete
jacket

Heavy concrete
saddle to hold
pipe down

7.3   The Alyeska Pipeline: oil from the North Slope

The Alyeska pipeline

In March 1968 at Prudhoe Bay on the shores of the Arctic Ocean a major oilfield was discovered 3,000 m below the tundra. The oil companies decided a pipeline was the appropriate way to move the oil from the North Slope of Alaska and formed a consortium (the Alyeska Pipeline) to build the pipeline. The 1,200 km of pipe had been delivered by the following year but construction didn't begin until 1974. The costliest private construction project in history ($7,700,000,000) had been delayed by a debate reflecting uncertainties about its design, route and ecological impact.

Conventional pipeline construction involves burial of the pipe in a trench. The Alaskan pipeline was going to be large, heavy and hot. With the diameter of 120 cm each metre of oil-filled pipe would weigh more than 900 kg and the oil leaves the ground at a temperature of 70–80°. Unfortunately 80% of Alaska is underlain by permafrost and in ice-saturated sediments such a warm buried pipe would thaw the permafrost to a radius of 9 m within five years. The pipe would soon lie in a trough of porridge-like mire—too thin to walk on and too thick to swim in! Its weight unsupported, the pipe would fracture and oil spillage result. An uninsulated pipe laid on the surface would still cause melt-out and erosion of the active layer as well as exposing the pipe to the $-70°C$ to $+30°C$ temperature ranges of interior Alaska.

'Wet' permafrost was only one problem. The line also traversed active fault zones and needed to cross 350 streams and rivers. The arctic environment lacks ecological diversity and because of this fragility is vulnerable to disruption. Any oil spillage would soon enter surface waters (being unable to seep away through the permafrost) and cause immediate and long-term damage to flora and fauna. In the harsh environment plant regeneration is extremely slow—the track of a bulldozer across tundra vegetation may still be fresh thirty years later. Such vehicle tracks and construction scars can cause localised thawing of the permafrost. Subsequent erosion causes silting in the streams and the destruction of salmon spawning grounds.

The construction options were conventional pipeline burial in stable areas and running the pipe above ground (either on an insulative gravel mat or supported clear of the ground on stilts) in unstable areas.

The industry-environment debate raged with claims and counter claims. At an early stage, for example, the USGS claimed 50% of the pipe should be above ground whilst the companies, conscious of cost, claimed 90% of the pipeline could be buried.

The first stage of the debate ended in the aftermath of the Yom Kippur war. During the Arab oil embargo the US Congress finally authorised the construction of the pipeline. Its route, problems and solutions are summarised in Fig. 7.3. For 680 km the pipe runs above ground on 78,000 supports frozen into the permafrost. The insulation-sheathed pipe is only fixed at intervals to allow for expansion and contraction and to allow adjustments to ground heaving and seismic disturbances. Designed to hopefully withstand the shock of an 8·5 (Richter scale) earthquake, 140 automatic cut-off valves would limit spillage to an average of 15,000 barrels in the event of fracture. Where caribou migration routes cross its line the pipe has been buried with refrigerated brine being pumped through smaller pipes in the trench to keep the permafrost stable.

By 1975 construction was in full swing. Each day's delay in production was costing the oil companies $2,600,000 so there was every reason for rapid progress! The pumping of oil didn't actually begin, however, until 1978. X-ray monitoring of joints had discovered 4,000 welding defects which had to be rectified before operations were allowed to begin.

The potential of Alaska and adjacent areas of Canada probably means that the Alyeska pipeline will be only the first of a network which will carry arctic oil and gas to the energy-hungry lower 48 states. The pipeline has been the focus of the largest and most expensive industry *versus* environment debate in history. Its major lessons will prove to be not only new designs and equipment but also better techniques and procedures for the monitoring of environments before, during and after such developments. Finally, and unfortunately beyond the focus of this book, the pipeline debate brought a greater awareness of the wide social costs of such developments.

CLIMATE AND MAN: URBAN CLIMATES

The 160 km² area shown in Fig. 7.4 lies at 51°N some 70 km east of the Rocky Mountains in Canada. It experiences a cold temperate climate, but with cooler summers than the prairies to the east because of its altitude and with warmer and more varied winters because of the ability of the Chinook (föhn) wind to raise temperatures from bitter cold to comparative warmth.

Imagine being at the RCMP post on a winter's afternoon in 1880. The wind is light, the sky partly cloudy. As the sun sets, long-wave radiation from the ground surfaces causes it to cool. Air close to the ground begins to fall in temperature and by midnight this cold denser air begins gliding downslope into the valley bottoms. This nocturnal cold air drainage produces pools of air as cool as −20°C, whilst a few hundred metres above on the prairie surface the temperature is −15°C. At dawn, when the sun rises, the ground surface absorbs the short-wave radiation, the air above it warms and the cold air pockets dissipate.

On a January night 86 years later it is a very different story. The area is now the home of 400,000 people. Grasses and trees have been replaced by concrete, asphalt, brick and tile—the city of Calgary. Figure. 7.5 shows the temperature distribution at 23.00 hours on a January night. The city has produced a 'heat island', 4°C warmer than the surrounding prairie and 7°C warmer than the cold air pocket in the Bow river valley to the west.

How has this happened? Calgary's homes, offices and factories emit radiant heat as energy escapes from their heating and lighting systems, cars circulate in the streets and the solar radiation stored from the daytime sun by the concrete and brick is re-radiated as long-wave radiation back to the atmosphere within the city. These processes combine to raise the night-time air temperatures of the urban area.

Figure 7.6 shows temperatures for three sites within the city, taken at four times during the day over the winter months. The daytime temperatures show only small differences. At night the story is different, warmer temperatures are recorded in the more densely built-up city centre. The kind of heat-island effect shown in Fig. 7.5 for a specific day and time is therefore detectable in *longer-term* records.

Such heat islands were recognised in the last century. Their shapes became clearer when measurements of air temperatures from moving cars became possible in the 1920's and more recently helicopters have begun to give us information on their three-dimensional structure. The sharp contrast in the thermal characteristics of the city and the surrounding countryside reflects the operation of five factors.

Firstly, there is *heat production* by man, low-level radiant heat from buildings and sensible heat from power station cooling waters and sewage effluent. In mid-latitudes during winter this heat production can total almost 50% of the incoming solar radiant energy, although it drops to about 15% during the summer.

Secondly, there are the changes in the *radiation balance* between the rural areas and the city. Most building materials have high thermal capacities and on a diurnal scale this means that solar radiant energy received during the day is stored in the brick and concrete and released during the night. The fabric of the city thus acts as a massive storage radiator. The absorption of solar radiation occurs on all surfaces of buildings and not just the roofs; a complex of tower blocks can in fact absorb six times as much radiation as agricultural land! The importance of this radiation-balance factor can be gauged from the fact that London's heat island is greater in summer, when there is more solar radiation and atmospheric heating by com-

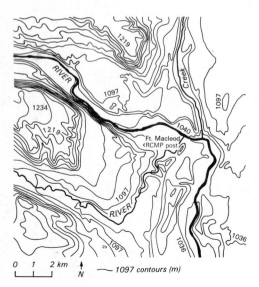

0   1   2 km   N   — 1097 *contours (m)*

7.4   Topography of the study area

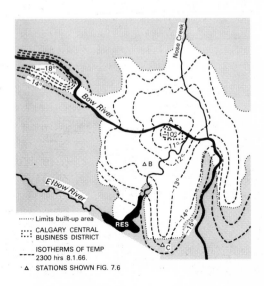

····· Limits built-up area
:::: CALGARY CENTRAL BUSINESS DISTRICT
---- ISOTHERMS OF TEMP 2300 hrs 8.1.66.
△ STATIONS SHOWN FIG. 7.6

7.5   Temperature distribution

OBSERVATION POINT	TIME OF OBSERVATION (hrs)			
	0800	1400	1600	2300
City centre (A)	−11	−6.5	−6.5	−10.5
Medium Density Residential (B)	−12.5	−7	−7.5	−12
Edge of built-up area (C)	−13	−7	−7.5	−12.5

(Mean temperatures for Dec, Jan and Feb 1965-6)

7.6   Calgary, city and country temperatures

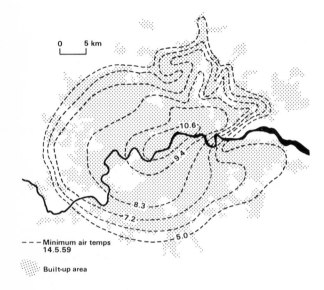

--- Minimum air temps
14.5.59

▦ Built-up area

7.8   A London heat island

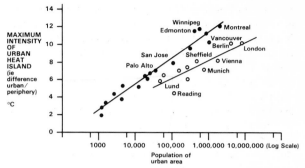

7.9   Heat islands and city size

bustion (the first factor above) is at a minimum.

The third factor is the effect of buildings on heat diffusion through their effects on *airflow*. The turbulent transfer of heat from the city is hindered by the reduction in windspeeds within the urban area. Fourthly, if the surface of the urban area is compared to the countryside there are fewer plants and smaller areas of open water. This means that less energy is used within the city for the *evaporation* and *transpiration* of water and more is available for heating the urban surfaces.

Finally, there are changes in the *composition of the atmosphere* in and over the city, a point to be mentioned again below. Smoke and haze although reducing incoming solar radiation have a blanketing effect during the night (atmospheric counter radiation—see Chapter 2) which raises air temperatures and contributes to the heat-island effect.

7.7   Calgary, view west from city centre along the Bow valley

The existence, intensity and shape of heat islands is influenced by a range of factors. High wind speeds break up the island. Ridges, plateaus and valleys, together with variations in the city surface (parkland, industry, high-rise buildings and different residential densities), determine its detailed shape. Figure 7.8 shows a London heat island. Although its margin follows the limit of the built-up area its centre is displaced north-eastwards. Chandler attributed this to two factors—south-westerly winds and closely packed housing of high thermal capacity at the centre of the island.

City size may also be a factor. Oke in a study in Eastern Canada discovered heat-island intensity was proportional to urban population. He then collated heat-island studies from a number of cities and his results are shown in Fig. 7.9. There is an interesting difference in the slope of the regression for the European and American cities—in other words, as you can see from inspecting Fig. 7.9, for cities of equal population size the European heat island is less intense than the American. This is somewhat of a paradox if density of the urban areas is considered the only factor. You might like to think about the differences in urban land use, vertical development and energy consumptions in the two groups of societies which might underpin this.

The urban heat island is only one manifestation of man's impact on the atmosphere. He has also modified its composition, a change which is most intense in urban areas although some of its effects, like CO_2 levels, are global in scale. This change has been achieved in two ways. Firstly by injecting additional particulate matter and aerosols (finely suspended matter) into the atmosphere. This is mainly smoke (carbon and tarry hydrocarbons) from the incomplete combustion of fossil fuels, although lead, aluminium and silica compounds have been added, too. The overall effect of this is to reduce sunlight transmission. Before the 1956 Clean Air Act, for example, many British cities lost 25–55% of their winter sunshine. Under inversion conditions the smoke concentrations became particularly intense, the classic London fog of 5–9th December 1952 reduced visibility to less than 10 m for 48 consecutive hours and caused 12,000 deaths.

The other modification is through the injection of gases. SO_2 from the sulphur content of fossil fuel is a serious pollutant—being soluble in water it attacks minerals readily. Figure 1 (in the Introduction) represented the Los Angeles' physical and cultural system.

ELEMENT		COMPARISON WITH RURAL AREA
Temperature	winter minima annual mean	+1—2°C +0.5—0.8°C
Radiation	total winter U/V sunshine hours	−15—20% −30% −5—15%
Wind	Annual mean vel. calms	−20—30% +5—20%
Cloudiness	cloud winter fog	+5—10% +100%
Precipitation	Total days with <5mm	+5—10% +10%
Atmospheric composition	Particulates CO_2 SO_2 N oxides CO hydrocarbons	x 3.7 x 2 x 200 x 10 x 200 x 20

7.10   Urban climatic conditions

In this topographic basin subjected to temperature inversions, 4,000,000 cars burn 30,000,000 litres of petrol a day. The unburnt petrol, highly reactive petro-chemical compounds and carbon monoxide in the vehicle exhausts combine to make a veritable 'witches brew' of daytime summer and autumn smog. Car exhausts, for example, emit NO. With hydrocarbons and sunlight this is turned into NO_2, a reaction which releases particulates (producing haze)—nature's inadequate way of cleansing the atmosphere. This reaction is accompanied by smell, eye irritation, plant damage and excessive ozone production.

Urban climates and pollution are interesting fields and the few paragraphs above have hardly scratched the surface. The overall effects of the urban area are summarised in Fig. 7.10 but the references should be used to follow this topic further.

THE FLOODING OF LONDON

The most expensive potential disaster facing Britain today is the flooding of London. Over a million people, 45 hospitals, 56 telephone exchanges and 50 underground stations are at risk in the 116 km² area which could be inundated by the tidal flooding of the Thames. The disruption to life, commerce and industry are almost immeasurable, with people trapped in their homes, sewage in the streets and parts of the underground system taking months to clean up.

The hazard of tidal flooding is caused by a combination of long- and short-term factors, one man-induced and two reflecting the operation of natural systems. In 1791 high tide levels reached 4·2 m OD at London Bridge, by 1881 4·9 m and by 1953 5·4 m. This rise reflects the gradual relative sinking of SE England (a glacial legacy) and the sinking of London itself on its clay bed as water is abstracted from the aquifers. Combined with this long-term relative sea-level rise is the surge threat. *Surges* are caused by travelling low-pressure systems—a pressure drop of 30 mb causing the sea surface to rise 30 cm. A potentially very unpleasant combination occurs when a depression passes to the north of Scotland and then moves southwards into the North Sea. Gale-force northerly winds on the westerly side of the depression accentuate the surge and drive it down the east coast, taking about eleven hours to travel from Aberdeen to the Thames estuary. The narrowing and shallowing basin of the southern North Sea increases the amplitude of the surge; a January 1978 surge, for example, 68 cm at the Tyne had risen to 1·72 m by the time it reached Southend. This process is magnified within the Thames estuary itself.

When such surges coincide with high water in the fortnightly tide cycle the scene is set for potential disaster. This was the situation in 1953 when a 2·75 m surge was added to high tides. Extensive areas of the East Coast were inundated, 300 lives were lost and 30,000 people evacuated from their homes. A point

made earlier in the book was that people's perception of hazard is stronger immediately after they have experienced one! The Waverley Committee appointed after this disaster called for the construction of coastal defences, a Thames flood barrier and a warning system. Nineteen years passed before a decision was approved to construct a barrier at Woolwich, its construction began in 1974 and it is scheduled for completion in 1983, thirty years after the disaster which focused attention on the need for it! Bank raising along the Thames has been undertaken to give some measure of interim protection combined with a flood-warning scheme. The actual barrier is a rising gate design which allows normal use of the river (Fig. 7.11).

One consequence of the 1953 East Coast flooding has been the development of the Storm Tide Warning Service at Bracknell which links meteorologists and hydrographers in a prediction function. The procedure basically involves monitoring a depression and its surge on its arrival off the north of Scotland. In addition to the normal weather reports Bracknell has a direct link with tide gauges from the Hebrides to the East Coast of England. If the actual surge is larger than predicted and the wind appears close to its worst possible direction then forecasting proceeds. This usually predicts surge amplitudes to within 15 cm but given the variables involved the predictions can be 60 cm out. During the late 1970's each year saw the issuing of 100 preliminary alerts (given 12 hours

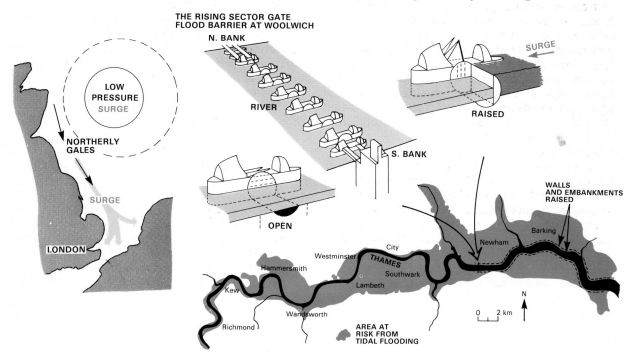

7.11   Tidal surge flooding

ahead), 20 confirmations and five actual danger warn-ings. How many of these should be allowed to filter through to the public is a big problem for the auth-orities: 'crying wolf' too often can cause apathy yet people should be aware of procedures if chaos and loss of life are to be minimised.

Until the barrier is complete London will be keeping its fingers crossed. Over the longer term, of course, the environmental systems continue to operate. One fore-cast is that the barrier will be trundling up and down ten times a year by the turn of the century!

CONCLUSION

The range of case-study detail relevant to a section with this theme has made it rather frustrating to write. Nonetheless it is hoped that the three examples in-troduced here, together with material in Chapters 4 and 5 illustrate the theme on interaction between man and environment.

The perfect environment is not the world of the peregrine falcon, neither is it the world of the cater-pillar tractor and cement-mixer! Cost–benefit analysis applied to decisions about resource development, pol-lution and hazard management can give some guide-lines for action. It is important, however, to remember that the 'costs' and 'benefits' reflect the value judge-ments of decision-makers and their priorities may well not be shared by all groups within society.

One justification for the study of earth science is the application of its knowledge in the service of society. Understanding the operation of natural systems and predicting the effects of change are thus basic goals. Nature, however, doesn't fall neatly into the fields of study which man has evolved so applied investigations of such problems invariably becomes interdisciplinary in nature. For questioning and concerned minds the interdisciplinary trail doesn't stop at the boundary of the earth sciences—the cultural, economic and politi-cal systems may be relevant, too. With the dawn of the plutonium age upon us the need for basic under-standing of natural systems, and an informed society able to question priorities and examine alternatives, has possibly never been greater.

Review Questions

1. Describe and account for the modifications to the atmosphere which take place in urban areas.

2. In the light of Alaskan experience what procedures do you think the Canadians could adopt in plan-ning the exploitation of oil and gas in the Mack-enzie Delta and Arctic Islands?
3. Why are the shores around the southern North Sea susceptible to tidal flooding?
4. Write an explanatory account of Dutch experience in constructing coastal defences and polder reclamation.
5. You have meteorological equipment (thermo-graphs and max/min thermometers) for ten fixed stations and two cars equipped with temperature-measuring equipment. For an urban area with which you are familar, describe and justify a pro-gramme of observations and analysis of the urban climate.
6. Many buildings (homes, schools, factories and offices) 'fight' the climate with inappropriate shapes, windows etc. and they use fossil fuels to heat and/or ventilate. For four environments (Arc-tic, mid-latitude temperate, the Persian Gulf and Equatorial) describe and comment on the possi-bilities for more environmentally sympathetic de-signs. (You may get some ideas from the traditional designs, materials, colours, etc., of these areas.)

Further Reading

Bennett, R. J., & Chorley, R. J., 1978, *Environmental Systems*, Methuen.
Brown, R. J. E., 1970, *Permafrost in Canada*, Uni-versity of Toronto Press.
Chandler, T. J., 1965, *The Climate of London*, Hut-chinson.
Cooke, R. U., & Doornkamp, J. C., 1974, *Geomor-phology in Environmental Management*, Oxford University Press.
Detwyler, T. R., 1971, *Man's Impact on Environment*, McGraw-Hill.
Legget, R. F., 1973, *Cities and Geology*, McGraw-Hill.
Oke, T. R., 1978, *Boundary Layer Climates*, Methuen.
Simmons, I. G., 1974, *The Ecology of Natural Resources*, Arnold.
Smith, K., 1975, *Principles of Applied Climatology*, McGraw-Hill.
Tank, R. W. (Ed.), 1976, *Focus on Environmental Geology*, Oxford University Press.
White, G., 1974, *Natural Hazards*, Oxford University Press.

Index